T0201106

A PRIMER ON STABLE ISOTOPES IN ECOLOGY

A Primer on Stable Isotopes in Ecology

M. Francesca Cotrufo

Department of Soil and Crop Sciences, Colorado State University, Fort Collins, CO, USA

Yamina Pressler

Natural Resources Management and Environmental Sciences, California Polytechnic State University, San Luis Obispo, CA, USA

OXFORD
UNIVERSITY PRESS

Great Clarendon Street, Oxford, OX2 6DP,
United Kingdom

Oxford University Press is a department of the University of Oxford.
It furthers the University's objective of excellence in research, scholarship,
and education by publishing worldwide. Oxford is a registered trade mark of
Oxford University Press in the UK and in certain other countries

Published in the United States of America by Oxford University Press
198 Madison Avenue, New York, NY 10016, United States of America

British Library Cataloguing in Publication Data
Data available

Library of Congress Control Number: 2023935116

ISBN 978–0–19–885449–4

DOI: 10.1093/oso/9780198854494.001.0001

Printed and bound by
CPI Group (UK) Ltd, Croydon, CR0 4YY

Preface

I became interested in isotopes during my graduate studies (1991–1994), when my advisor, Dr. Phil Ineson, and I were brainstorming about how to quantify changes in plant carbon (C) inputs belowground under elevated atmospheric carbon dioxide. We concluded that only quantitative isotope tracing could answer that question. Since then, I have been an isotopes fan, always in search of learning more about them! Visiting Dr. Jim Ehleringer's laboratory and participating in the Stable Isotope Ecology Course at Utah State University made me into an "Isotopeer" and further deepened my fascination for stable isotopes.

I see stable isotopes as a powerful tool to advance our understanding of soil biogeochemistry and find joy in sharing my research passions with students. I established and coordinated a graduate degree program in "Development and application of isotopic methodologies to environmental science research" at my former University of Campania, Italy. Once at Colorado State University, I designed and continue to teach a graduate course in "Terrestrial Ecosystem Isotope Ecology." However, in all these years, my students and I suffered from the lack of a textbook that clearly covered the basic principles for the correct application of stable isotopes in ecological research. This book was inspired by this sentiment and aims to fill this gap.

Embarking on a textbook-writing project, when I was already overcommitted by the research, teaching, and service tasks of a faculty member, took courage and a lot of enthusiasm. For someone like me, not expert in pedagogy and not a native English speaker, it also required help from a talented writer with pedagogy skills. This book would not exist if Yamina Pressler, the most talented writer and pedagogy expert in my research group at the time, hadn't responded with the most enthusiastic "Yes, let's do it!" when I shared the idea with her. Together, we began crafting a textbook tailored specifically to ecologists and biogeochemists looking to understand and apply stable isotopes in their work.

In the past few decades, the field of ecology and in particular terrestrial ecosystem ecology has made huge advancements thanks to stable isotopes. Today, you will find presentations using isotope techniques in every ecological conference, and many publications that apply stable isotopes to answer questions in ecology and biogeochemistry. Ecologists need to understand the principles of stable isotopes to fully appreciate many studies in their discipline. Ecologists also need to be aware of isotopic approaches to enrich their "toolbox" for further advancing our discipline. The increasing number of short, intensive, and conventional graduate courses on stable isotope ecology all over the world testifies to this. However, anyone interested in stable isotope ecology, in particular for applications in terrestrial ecosystems, has to navigate a myriad of scientific publications and book chapters, each offering a piece of the puzzle. Sometimes, researchers apply isotopic techniques with a very limited understanding of the underlying theory, leading to erroneous conclusions. We believe these challenges emerge from the lack of an approachable book to learn from and refer to.

We wrote this book to address these needs and provide a concise, hopefully enjoyable, and solid background reading for anyone interested in acquiring state-of-the-art theory and practical knowledge for the application of stable isotopes in ecology. This book is designed for students and scientists from different backgrounds who share the common interest of exploring stable isotope methods in their research, or simply would like to gain a more in-depth and complete knowledge of stable isotopes. The text is designed to guide readers to think "isotopically" to better understand research conducted using stable isotopes. The book also provides basic practical skills to apply isotope methods in ecological research. Readers will learn about stable isotopes and their notation, isotope fractionation, mixing, heavy isotope enrichment, and quantification methods by mass spectrometry and laser spectroscopy.

Even writing a relatively short book like this requires help and support along the way. A big thanks goes to Dr. Jesse Nippert and Dr. Roland Bol for reviewing the text and providing very useful and enthusiastic comments; Dr. Jim Ehleringer and Dr. Phil Ineson for sharing slides and photos; and Dr. Carsten Müller, Dr. Jennifer Pett-Ridge, and Dr. Livio Gianfrani for providing subject-specific suggestions and sharing publications. I am very grateful to all the students in my laboratory and isotope classes, who kept me inspired through all these years, and sharpened and shaped the way in which I think about and present this material.

M. Francesca Cotrufo
August 31, 2022
Fort Collins, CO

Contents

List of Abbreviations

APE	atom percent excess
C	carbon
CAM	crassulacean acid metabolism
CDT	Canyon Diablo Troilite
EA	elemental analyzer
GC-c-IRMS	gas chromatography–combustion–mass spectrometry
H	hydrogen
IAEA	International Atomic Energy Agency
IRMS	isotope ratio mass spectrometry
LC	liquid chromatography
NanoSIMS	nano secondary ion mass spectrometry
NIST	National Institute of Standards and Technology
N	nitrogen
NMR	nuclear magnetic resonance
O	oxygen
P	phosphorus
PDB	PeeDee Belemnite
PEP	phosphoenolpyruvate
PLFA	phospholipid fatty acid
qSIP	quantitative stable isotope probing
RuBisCO	ribulose 1,5-bisphosphate carboxylase/oxygenase
S	sulfur
SI	International System of Units
SIP	stable isotope probing
SMOW	Standard Mean Ocean Water
SOM	soil organic matter
TDL	tunable diode laser
VSMOW	Vienna Standard Mean Ocean Water Standard

1

Stable Isotopes as a Tool for Ecologists

1.1 Introduction

Ecologists are transdisciplinary scientists, integrating biology, physics, chemistry, physiology, biogeochemistry, and several other disciplines in their studies. Their tool kit is thus vast and very diverse, spanning from simple observations to state-of-the-art sophisticated instrumentation. Stable isotopes offer a powerful tool of study for any subdiscipline of ecology.

Ecosystem, plant, soil, microbial, and animal ecologists can all use stable isotopes to address their questions. How long does it take for a carbon dioxide (CO_2) molecule to travel from the atmosphere through the plant, into the soil, and be respired back to the atmosphere? How efficiently does a plant use water, or where do plants source water from? How much of the organic matter of a soil in a savanna-forest tropical ecotone is derived from the savanna grasses, which use a C4 photosynthetic pathway, and how much from the trees, which use a C3 photosynthetic pathway (see Section 3.6)? Which microbial group has consumed a specific substrate? What is the migration route of a snow goose or a whale? What are the trophic relationships in a food web? These, and many more, are questions that can be answered using stable isotopes at natural abundance or after enrichment manipulations.

Stable isotopes are a powerful tool in ecology because they are "one of nature's ecological recorders" (West et al. 2006) as they integrate, indicate, record, and trace fundamental ecological processes across a broad range of temporal and spatial scales (Figure 1.1; Tu et al. 2007).

1.2 Stable isotopes integrate ecological processes in space and time

The isotope ratio of any biogeochemical pool represents a temporal integration of the processes responsible for fluxes into and out of a pool. The timescale of this integration depends on the element turnover rate within the pool in question. Because of this, stable

A Primer on Stable Isotopes in Ecology. M. Francesca Cotrufo and Yamina Pressler, Oxford University Press.
 DOI: 10.1093/oso/9780198854494.003.0001

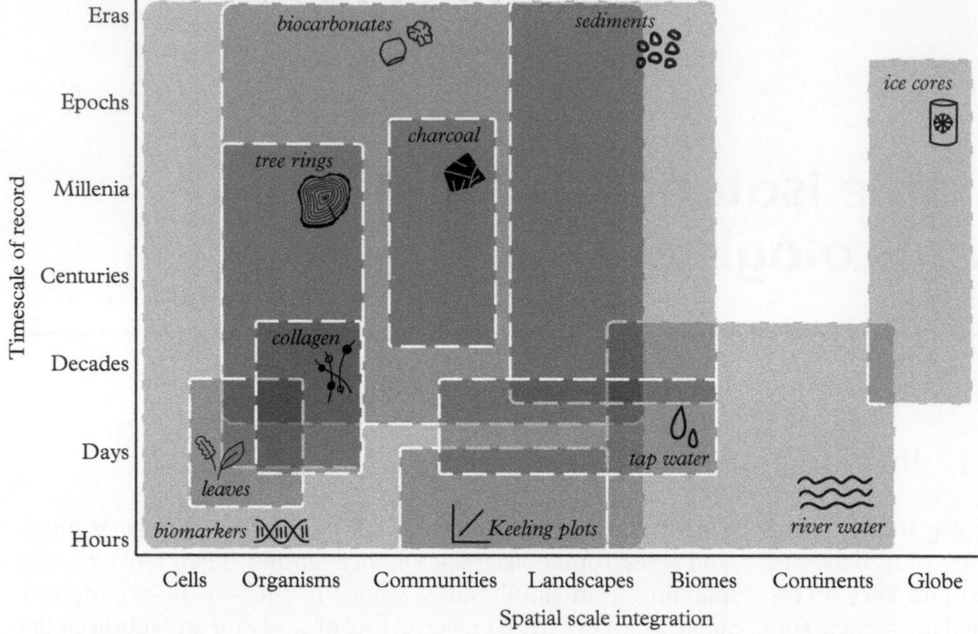

Figure 1.1 *Stable isotope analyses of a variety of samples can cover wide spatial and temporal scales. Spatial and temporal scales refer to the spatial and temporal integration, respectively, of a sample used for isotope analyses. For example, a sample of ice from an ice core integrates spatially across the entire globe, and temporally from a yearly resolution to hundreds of thousands of years.*
Modified from Tu et al. (2007).

isotopes are a powerful tool to study turnover rates of biogeochemical pools. This property has been extensively used, for example, to determine the turnover rates of soil organic carbon pools as well as of specific compounds, both at natural abundance after land-use conversion from C3 to C4 vegetation or vice versa (e.g., Balesdent et al. 1987; Del Galdo et al. 2003) or through heavy isotope enrichments (e.g., Wiesenberg et al. 2008).

Similarly, the temporal integration of stable isotopes in animal and plant tissues provides information on physiological and ecological processes on the landscape. For example, the nitrogen-15 (^{15}N) isotopic composition of tissues of migratory birds can inform on birds' state of fatigue or starvation (Hobson et al. 2010), or the carbon-13 (^{13}C) abundance in the roots of C3 and C4 plants living in proximity can inform us on the existence of symbiotic relationships through common mycorrhizal networks (Watkins et al. 1996).

For well-mixed biogeochemical pools, such as the atmosphere, the stable isotope composition integrates source inputs to the pool, extending over large spatial and temporal scales. For example, since the industrial revolution in the second half of the eighteenth century, anthropogenic emissions of CO_2 from the combustion of fossil fuels have contributed ^{13}C-depleted CO_2 to the Earth's atmosphere. Over time, these have mixed and integrated with the CO_2 derived from natural biological and geological processes, resulting in a continuous decrease of delta (δ; see Section 2.3) ^{13}C of atmospheric CO_2 at a global scale, known as the *Suess effect* (Keeling 1979).

1.3 Stable isotopes indicate the presence and magnitude of ecological processes

Through fractionation (a process that we will learn about in Chapter 3), ecological processes produce distinctive stable isotope compositions in plant, animal, and microbial tissues, and other biogeochemical pools. Thus, the stable isotope composition of an element in a pool reflects the presence and magnitude of such processes. Let's look at a couple of examples to understand this concept.

Photosynthesis in C3 plants (see Figure 3.6) fractionates against ^{13}C with the level of fractionation depending on the degree of opening of leaf stomata. Since stomatal opening also determines water losses, photosynthetic ^{13}C fractionation is linearly related to the amount of water molecules a C3 plant loses to acquire a molecule of CO_2, also known as water use efficiency. For this reason, the natural abundance of ^{13}C in the leaf of C3 plants indicates CO_2 uptake and transpiration processes and can be used to determine the magnitude of water use efficiency (Ehleringer et al. 1993).

Nitrogen is lost from soils via leaching, denitrification, or volatilization, which all fractionate against the heavy isotope ^{15}N (i.e., the ^{14}N is preferentially lost). Thus, ^{15}N accumulation in soil typically indicates the presence of N loss processes, and the higher the ^{15}N accumulation, the higher the magnitude of those losses. For example, Frank and Evans (1997) quantified the ^{15}N abundance in soils at Yellowstone National Park, USA, to better understand ecosystem N dynamics in grazed and ungrazed sites. They observed ^{15}N enrichments in native grazers' urine and dung patches indicating higher N loss processes from these patches than control ungrazed soils (Frank and Evans 1997).

1.4 Stable isotopes record bio-physical responses to changing environments

When substrates accumulate incrementally over time, the stable isotope composition of the substrate can be used as a record of ecological responses or as a proxy for environmental changes across the time covered by the sample (Dawson and Siegwolf 2011). Such incremental accumulation occurs in tree rings, animal hair, ice cores, soil cores, and other ecological materials.

Most trees deposit annual rings in their trunk. Analyzing the ^{13}C, oxygen-18 (^{18}O), or hydrogen-2 (^{2}H, or deuterium, D) natural abundance of each ring bulk tissue or celluloses enables the reconstruction of temporal records of a variety of climatic variables, from local to regional scales (McCarroll and Loader 2004). Stable isotopes in tree rings are also a powerful tool to record tree physiological responses to natural or anthropogenic events such as volcanic eruptions (Battipaglia et al. 2007) or fires (Beghin et al. 2011).

In terrestrial mammals, hair grows over time. Thus, their isotopic composition can be used to determine changes in these animals' diet at the time resolution of the rate of formation of the length of hair increment measured. For example, Cerling et al. (2006) reconstructed stable isotope chronologies from hair of elephants in Northern Kenya to determine their dietary changes between browsing on C3 vegetation and feeding on C4 savanna grasses.

1.5 Stable isotopes source and trace movement of key elements, substrates, and organisms

Due to fractionation, the stable isotope composition of element pools within and among ecosystems often differ. Because of this, and the conservation of mass, we can use isotopic mixing models to determine the relative contribution of source pools with contrasting isotopic fingerprints to a mixture pool (as we will learn about in Chapter 4). The use of stable isotopes to partition source contributions is widely used in ecology. Examples include the reconstruction of food sources in animal diets (Phillips and Koch 2002; Tykot 2004), the partitioning of water or N sources in plants (Dawson et al. 2002), and the quantification of the relative contribution of different vegetation inputs to the soil organic matter pool (Balesdent et al. 1987).

Additionally, since the source of an essential element acquired by an organism can be traced using stable isotope ratios, we can track the origin of substrates and movements of organisms across landscapes and continents. Geographical patterns of the natural abundance of stable isotopes are known as isoscapes (Bowen 2010). Isoscapes are commonly used to track the routes of migratory animals (Hobson et al. 2010) and have many other applications.

As an ecologist, you will come across many applications of stable isotopes in the literature and may consider applying the approach to your own research questions. This book is designed to provide ecologists with the basic principles and concepts required to understand, interpret, and use stable isotopes in their research. While we present a few examples for applications of stable isotopes in ecology, mostly from the field of biogeochemistry and soil ecology which are our fields of research, we do not intend to describe all ecological applications of stable isotopes. We encourage readers to explore the diverse ecological applications of stable isotopes in previously published books (Ehleringer et al. 1993; Unkovich et al. 2001; Flanagan et al. 2004; Faure and Mensing 2005; Fry 2006; Michener and Lajtha 2008; Dawson and Siegwolf 2011; Rundel et al. 2012) and published subject-specific research papers in areas of interest.

References

Balesdent, J., A. Mariotti and B. Guillet (1987). "Natural ^{13}C abundance as a tracer for studies of soil organic matter dynamics." *Soil Biology and Biochemistry* **19**(1): 25–30.

Battipaglia, G., P. Cherubini, M. Saurer, R. T. W. Siegwolf, S. Strumia and M. F. Cotrufo (2007). "Volcanic explosive eruptions of the Vesuvio decrease tree-ring growth but not photosynthetic rates in the surrounding forests." *Global Change Biology* **13**(6): 1122–1137.

Beghin, R., P. Cherubini, G. Battipaglia, R. Siegwolf, M. Saurer and G. Bovio (2011). "Tree-ring growth and stable isotopes (^{13}C and ^{15}N) detect effects of wildfires on tree physiological processes in Pinus sylvestris L." *Trees* **25**(4): 627–636.

Bowen, G. J. (2010). "Isoscapes: spatial pattern in isotopic biogeochemistry." *Annual Review of Earth and Planetary Sciences* **38**(1): 161–187.

Cerling, T. E., G. Wittemyer, H. B. Rasmussen, F. Vollrath, C. E. Cerling, T. J. Robinson and I. Douglas-Hamilton (2006). "Stable isotopes in elephant hair document migration patterns and diet changes." *Proceedings of the National Academy of Sciences USA* **103**(2): 371–373.

Dawson, T. E., S. Mambelli, A. H. Plamboeck, P. H. Templer and K. P. Tu (2002). "Stable Isotopes in Plant Ecology." *Annual Review of Ecology and Systematics* **33**: 507–559.

Dawson, T. E. and R. Siegwolf (2011). *Stable isotopes as indicators of ecological change*. London, Academic Press.

Del Galdo, I., J. Six, A. Peressotti and M. F. Cotrufo (2003). "Assessing the impact of land-use change on soil C sequestration in agricultural soils by means of organic matter fractionation and stable C isotopes." *Global Change Biology* **9**(8): 1204–1213.

Ehleringer, J. R., A. E. Hall and G. D. Farquhar, eds. (1993). *Stable isotopes and plant carbon–water relations* (Physiological Ecology series). San Diego, CA, Academic Press.

Faure, G. and T. M. Mensing (2005). *Isotopes: principles and applications*. Hoboken, NJ, John Wiley & Sons, Inc.

Flanagan, L. B., J. R. Ehleringer and D. E. Pataki, eds. (2004). *Stable isotopes and biosphere-atmosphere interactions*. San Diego, CA, Elsevier.

Frank, D. A. and R. D. Evans (1997). "Effects of native grazers on grassland N cycling in Yellowstone National Park." *Ecology* **78**(7): 2238–2248.

Fry, B. (2006). *Stable isotope ecology*. New York, Springer.

Hobson, K. A., R. Barnett-Johnson and T. Cerling (2010). Using isoscapes to track animal migration. In: J. B. West, G. J. Bowen, T. E. Dawson and K. P. Tu, eds. *Isoscapes: understanding movement, pattern, and process on Earth through isotope mapping*. Dordrecht, Springer Netherlands: 273–298.

Keeling, C. D. (1979). "The Suess effect: ^{13}Carbon-^{14}Carbon interrelations." *Environment International* **2**(4): 229–300.

McCarroll, D. and N. J. Loader (2004). "Stable isotopes in tree rings." *Quaternary Science Reviews* **23**(7): 771–801.

Michener, R. and K. Lajtha (2008). *Stable isotopes in ecology and environmental science*. Hoboken, NJ, John Wiley & Sons.

Phillips, D. L. and P. L. Koch (2002). "Incorporating concentration dependence in stable isotope mixing models." *Oecologia* **130**(1): 114–125.

Rundel, P. W., J. R. Ehleringer and K. A. Nagy (2012). *Stable isotopes in ecological research*. New York, Springer Science & Business Media.

Tu, K. P., G. J. Bowen, D. Hemming, A. Kahmen, A. Knohl, C. T. Lai and C. Werner (2007). "Stable isotopes as indicators, tracers, and recorders of ecological change: synthesis and outlook." *Terrestrial Ecology* **1**: 399–405.

Tykot, R. H. (2004). Stable isotopes and diet: you are what you eat. In: M. Martini, M. Milazzo, and M. Piacentini, eds. *Physics methods in archaeometry*. Amsterdam: IOS Press: 433–444.

Unkovich, M. J., J. S. Pate, A. McNeill and J. Gibbs (2001). *Stable isotope techniques in the study of biological processes and functioning of ecosystems*. New York, Springer Science & Business Media.

Watkins, N. K., A. H. Fitter, J. D. Graves and D. Robinson (1996). "Carbon transfer between C3 and C4 plants linked by a common mycorrhizal network, quantified using stable carbon isotopes." *Soil Biology and Biochemistry* **28**(4): 471–477.

West, J. B., G. J. Bowen, T. E. Cerling and J. R. Ehleringer (2006). "Stable isotopes as one of nature's ecological recorders." *Trends in Ecology & Evolution* **21**(7): 408–414.

Wiesenberg, G. L. B., J. Schwarzbauer, M. W. I. Schmidt and L. Schwark (2008). "Plant and soil lipid modification under elevated atmospheric CO_2 conditions: II. Stable carbon isotopic values ($\delta^{13}C$) and turnover." *Organic Geochemistry* **39**(1): 103–117.

2

Stable Isotopes, Notations, and Standards

2.1 What is a stable isotope?

The material world as we know it is made up of a relatively low number of chemical elements whose properties are defined by the number of protons in their atomic nuclei, which we refer to as the *atomic number* (Z). In 1869, the Russian chemist Dmitri Mendeleev proposed to arrange the chemical elements on the basis of their atomic number in the first published version of what we know today as the periodic table of the elements (Figure 2.1).

Nuclides are nuclei with different configurations. A family of nuclides having the same number of protons, and therefore being of the same element, reside in the same place of the periodic table of elements. They are called *isotopes*. Isotopes vary in the number of *neutrons* (N) and therefore in their total *atomic mass* (A), where $A = N + Z$.

The word isotope comes from the phrase "*iso topos*" or "same place" in the ancient Greek language. In 1913, Frederick Soddy, a British physical chemist, was the first to define isotopes, by stating that "certain elements exist in two or more forms which have different atomic weights but are indistinguishable chemically." For this, he won the Nobel Prize in Chemistry in 1922 (Soddy 1923).

Let's explore this concept with an example. Carbon (C) atoms have six protons in their nucleus (i.e., an atomic number, Z, of 6) which positions them in a specific place on the periodic table. Carbon atoms, however, can exist with a different number of neutrons. The large majority of them (98.9%) have a number of neutrons equal to the number of protons ($N = Z = 6$), and therefore a total atomic mass of 12. Far fewer of them (1.1%) have an additional neutron in their nucleus ($N = 7, A = 13$). Finally, an even smaller number of C atoms ($<10^{-12}$%) have yet one more neutron ($N = 8, A = 14$) which makes them radioactive. We call C atoms with different atomic masses C isotopes and identify them by adding their atomic mass as a superscript to the left of the element symbol, so that ^{12}C, ^{13}C, and ^{14}C represent C atoms of atomic mass 12, 13, and 14, respectively (Figure 2.2).

A Primer on Stable Isotopes in Ecology. M. Francesca Cotrufo and Yamina Pressler, Oxford University Press.
 DOI: 10.1093/oso/9780198854494.003.0002

1 IA	2 IIA	3 IIIB	4 IVB	5 VB	6 VIB	7 VIIB	8 VIIIB	9 VIIIB	10 VIIIB	11 IB	12 IIB	13 IIIA	14 IVA	15 VA	16 VIA	17 VIIA	18 VIIIA
1 H Hydrogen 1.008																	2 He Helium 4.003
3 Li Lithium 6.941	4 Be Beryllium 9.012											5 B Boron 10.81	6 C Carbon 12.01	7 N Nitrogen 14.01	8 O Oxygen 16.00	9 F Fluorine 19.00	10 Ne Neon 20.18
11 Na Sodium 22.99	12 Mg Magnesium 24.31											13 Al Aluminium 26.98	14 Si Silicon 28.09	15 P Phosphorus 30.97	16 S Sulfur 32.07	17 Cl Chlorine 35.45	18 Ar Argon 39.95
19 K Potassium 39.10	20 Ca Calcium 40.08	21 Sc Scandium 44.95	22 Ti Titanium 47.86	23 V Vanadium 50.94	24 Cr Chromium 52.00	25 Mn Manganese 54.94	26 Fe Iron 55.85	27 Co Cobalt 58.93	28 Ni Nickel 58.69	29 Cu Copper 63.55	30 Zn Zinc 65.38	31 Ga Gallium 69.72	32 Ge Germanium 72.61	33 As Arsenic 74.92	34 Se Selenium 78.96	35 Br Bromine 79.90	36 Kr Krypton 83.80
37 Rb Rubidium 85.47	38 Sr Strontium 87.62	39 Y Yttrium 88.91	40 Zr Zirconium 91.22	41 Nb Niobium 92.91	42 Mo Molybdenum 95.94	43 Tc Technetium 98	44 Ru Ruthenium 101.1	45 Rh Rhodium 102.9	46 Pd Palladium 106.4	47 Ag Silver 107.9	48 Cd Cadmium 112.4	49 In Indium 114.8	50 Sn Tin 118.7	51 Sb Antimony 121.8	52 Te Tellurium 127.6	53 I Iodine 126.9	54 Xe Xenon 131.3
55 Cs Cesium 132.9	56 Ba Barium 137.3	57 – 71 Lanthanides	72 Hf Hafnium 178.5	73 Ta Tantalum 180.9	74 W Tungsten 183.8	75 Re Rhenium 186.2	76 Os Osmium 190.2	77 Ir Iridium 192.2	78 Pt Platinum 195.1	79 Au Gold 197.0	80 Hg Mercury 200.6	81 Tl Thallium 204.4	82 Pb Lead 207.2	83 Bi Bismuth 209.0	84 Po Polonium 209	85 At Astatine 210	86 Rn Radon 222
87 Fr Francium 223	88 Ra Radium 226	89 – 103 Actinides	104 Rf Ruther fordium 261	105 Db Dubnium 262	106 Sg Seaborgium 266	107 Bh Bohrium 264	108 Hs Hassium 277	109 Mt Meitnerium 268	110 Ds Darmstadtium 281	111 Rg Roentgenium 272	112 Cn Conipernicium 285	113 Nh Nihonium	114 Fl Flerovium	115 Mc Moscovium	116 Lv Livermorium	117 Ts Tennessine	118 Cg Oganesson

57 La Lanthanum 138.9	58 Ce Cerium 140.1	59 Pr Praseodymium 140.9	60 Nd Neodymium 144.2	61 Pm Promethium 145	62 Sm Samarium 150.4	63 Eu Europium 152.0	64 Gd Gadolinium 157.3	65 Tb Terbium 158.9	66 Dy Dysprosium 162.5	67 Ho Holmium 164.9	68 Er Erbium 167.3	69 Tm Thulium 168.9	70 Yb Ytterbium 173.0	71 Lu Lutetium 175.0
89 Ac Actinium 227	90 Th Thorium 232.0	91 Pa Protactinium 231.0	92 U Uranium 238.0	93 Np Neptunium 237	94 Pu Plutonium 244	95 Am Americium 243	96 Cm Curium 247	97 Bk Berkelium 247	98 Cf Californium 251	99 Es Einsteinium 252	100 Fm Fermium 257	101 Md Mendelevium 258	102 No Nobelium 259	103 Lr Lawrencium 262

Figure 2.1 *The modern periodic table of elements.*

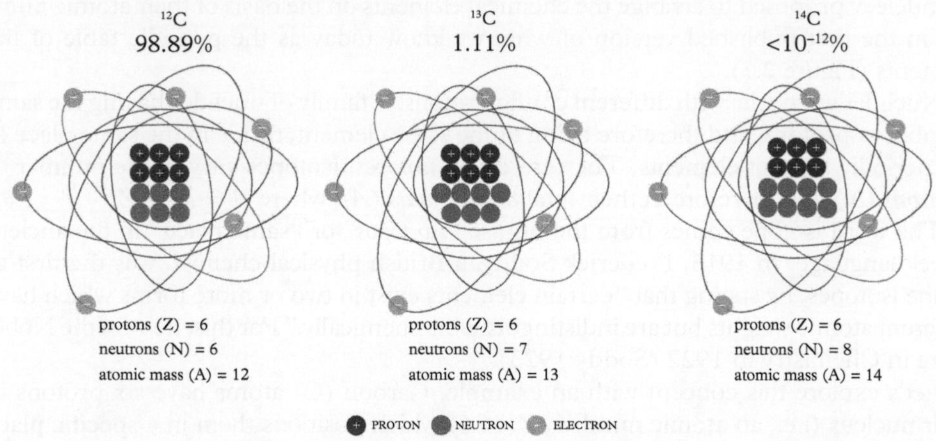

Figure 2.2 *Conceptual diagram of ^{12}C, ^{13}C, and ^{14}C isotopes demonstrating the relationship between protons (Z), neutrons (N), and atomic mass (A). Protons are shown in dark gray, neutrons in medium gray, and electrons in light gray. ^{12}C is more abundant (98.89% of Earth's natural C isotopes) than ^{13}C (1.11% natural abundance) and ^{14}C ($<10^{-12}$% natural abundance).*

Many elements have multiple isotopes. In addition to C mentioned above, nitrogen (N), oxygen (O), hydrogen (H), and sulfur (S) are some of the most studied isotopes in ecology and biogeochemistry. Therefore, we focus on them in this book. Nitrogen atoms exist as ^{14}N ($N = Z = 7$, $A = 14$) and ^{15}N ($Z = 7$, $N = 8$, $A = 15$). Oxygen atoms exist as ^{16}O ($N = Z = 8$, $A = 16$), ^{17}O ($Z = 8$, $N = 9$, $A = 17$),

Table 2.1 *Natural abundances of stable isotopes commonly used in ecology.*

Element	Stable isotope	Natural abundance (%)
H	^{1}H	99.984
	^{2}H or D	0.016
C	^{12}C	98.89
	^{13}C	1.11
N	^{14}N	99.64
	^{15}N	0.36
O	^{16}O	99.76
	^{18}O	0.20
S	^{32}S	95.02
	^{33}S	0.75
	^{34}S	4.21
	^{36}S	0.02

and ^{18}O ($Z = 8, N = 10, A = 18$). Hydrogen atoms exist as ^{1}H ($Z = 1, N = 0, A = 1$), ^{2}H or deuterium (D; $N = Z = 1, A = 2$), and ^{3}H or tritium ($Z = 1, N = 2, A = 3$), which is radioactive. Sulfur atoms exist under many isotopic configurations, four of which are stable: ^{32}S ($N = Z = 16, A = 32$), ^{33}S ($Z = 16, N = 17, A = 33$), ^{34}S ($Z = 16, N = 18, A = 34$), and ^{36}S ($Z = 16, N = 20, A = 36$). For each element one isotope is overwhelmingly more abundant on Earth than the others (Table 2.1). In general, the lightest isotope is more abundant than the heavier isotopes (Table 2.1).

Since isotopes are hidden within the periodic table of elements, we can visualize the variation in isotopes in a nuclide chart (Figure 2.3). In the nuclide chart, we can also distinguish families of nuclides with the same number of neutrons but varying numbers of protons, called *isotones*, and families of nuclides with the same atomic mass, called *isobars* (Figure 2.3).

By definition, isotopes of one element have different neutron:proton ratios, and therefore different nuclear stability. *When the number of neutrons is such to neutralize the repulsive forces of the protons, the isotope is stable.* That is, the nuclear configuration of a stable isotope will not change over time. By contrast, an imbalance between protons and neutrons results in an unstable or radioactive isotope, which will decay to a more energetically favorable nuclear configuration over time. Interestingly, of the 2500 known nuclides only 270 are stable. Radioactive isotopes, such as ^{14}C and phosphorus-32 (^{32}P) and -33 (^{33}P) are also used in ecology. In this book, we will only discuss stable isotopes.

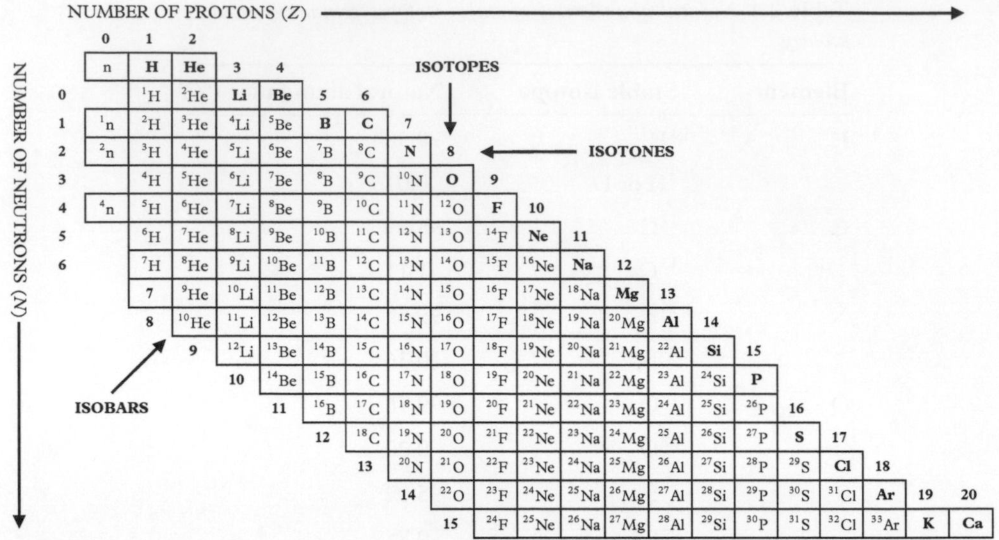

Figure 2.3 *Nuclide chart showing the relationship between isotopes, isotones, and isobars. Isotopes are nuclides with varying numbers of neutrons (N). Isotones are nuclides with varying numbers of protons (Z). Isobars are nuclides with the same atomic mass. Note that the chart is used to exemplify the relationship between isotopes, isotones, and isobars, but not all of the nuclides in the chart exist in nature.*

2.2 Isotopocules

Isotopes are nuclides of the same element. Isotopes have, by definition, the same chemical properties determined by the number of protons in the nucleus. Thus, different isotopes form the same molecules. Therefore, molecules with differing isotopic compositions or with the same isotopic composition, but different isotopic substitutions (i.e., different position of the heavy isotope in the molecule) both exist. Molecules with different isotopic compositions are defined as *isotopologues* and have different masses. Molecules with different isotope substitutions are defined as *isotopomers* and have the same mass. Together, isotopologues and isotopomers of the same molecule are termed *isotopocules*. However, because heavy isotopes form stronger chemical bonds than lighter isotopes, and bond strength affects molecular energy differently depending on the isotope position in the molecule, all isotopocules have different molecular energies (Box 2.1). These different molecular energies have widespread consequences for the biogeochemical cycling of stable isotopes in the environment—most notably, isotope fractionation. We will revisit isotopocules in the context of isotope fractionation in Chapter 3.

Box 2.1 Isotopologues, isotopomers, and together, isotopocules

Ecologists leverage the chemical properties of stable isotopes in molecules to understand, measure, and manipulate biogeochemical processes in nature. Understanding the isotopic composition of a molecule is a powerful tool for gaining insight into the dynamics of that molecule, and its components, in the Earth system. Therefore, we must have a clear understanding of how stable isotopes are arranged within molecules. See Table 2.2.

Table 2.2 *Definitions of key terms.*

Term	Definition
Isotopologues	Molecules with different isotopic composition, and therefore different masses
Isotopomers	Molecules with different isotopic substitutions, and therefore the same masses
Isotopocules	Isotopologues and isotopomers of the same molecule

Let's consider the molecule CO_2 as an example. Isotopic substitution can occur in both the C atoms and the O atoms within a CO_2 molecule. As a result, there are 12 isotopologues of CO_2, of which $^{12}C^{16}O_2$ and $^{13}C^{16}O_2$ are by far the most abundant (Table 2.3). Notice that in $^{13}C^{16}O_2$, the lighter and more abundant isotope (^{12}C) is replaced with the heavier and less abundant isotope (^{13}C). While much less abundant, ^{17}O and ^{18}O isotopes will also replace ^{16}O in some CO_2 isotopologues (Table 2.3).

Table 2.3 *Abundances and molecular energies of isotopocules of molecules relevant to ecology and biogeochemistry (isotopomers are indicated in parentheses).*

Molecule	Isotopocule	Abundance	Molar mass (g mol^{-1})
H_2O	$H_2{}^{16}O$	0.997317	18.010565
	$H_2{}^{18}O$	0.002000	20.014811
	$H_2{}^{17}O$	3.718840×10^{-4}	19.01478
	$HD^{16}O$	3.106930×10^{-4}	19.01674
	$HD^{18}O$	6.230030×10^{-7}	21.020985
	$HD^{17}O$	1.158530×10^{-7}	20.020956
	$D_2{}^{16}O$	2.419700×10^{-8}	20.022915

(*Continued*)

Table 2.3 *Continued*

Molecule	Isotopocule	Abundance	Molar mass (g mol^{-1})
CO_2	$^{12}C^{16}O_2$	0.984204	43.98983
	$^{13}C^{16}O_2$	0.011057	44.993185
	$^{16}O^{12}C^{18}O$	0.003947	45.994076
	$^{16}O^{12}C^{17}O$	7.339890×10^{-4}	44.994045
	$^{16}O^{13}C^{18}O$	4.434460×10^{-5}	46.997431
	$^{16}O^{13}C^{17}O$	8.246230×10^{-6}	45.9974
	$^{12}C^{18}O_2$	3.957340×10^{-6}	47.998322
	$^{17}O^{12}C^{18}O$	1.471800×10^{-6}	46.998291
	$^{12}C^{17}O_2$	1.368470×10^{-7}	45.998262
	$^{13}C^{18}O_2$	4.446000×10^{-8}	49.001675
	$^{18}O^{13}C^{17}O$	1.653540×10^{-8}	48.001646
	$^{13}C^{17}O_2$	1.537500×10^{-9}	47.0016182378
CH_4	$^{12}CH_4$	0.988274	16.0313
	$^{13}CH_4$	0.011103	17.034655
	$^{12}CH_3D$	6.157510×10^{-4}	17.037475
	$^{13}CH_3D$	6.917850×10^{-6}	18.04083
N_2	$^{14}N_2$	0.992687	28.006148
	$^{14}N^{15}N$	0.007478	29.003182
N_2O	$^{14}N_2{}^{16}O$	0.990333	44.001062
	$^{14}N^{15}N^{16}O$ ($^{15}N\alpha$-N_2O)	0.003641	44.998096
	$^{15}N^{14}N^{16}O$ ($^{15}N\beta$-N_2O)	0.003641	44.998096
	$^{14}N_2{}^{18}O$	0.001986	46.005308
	$^{14}N_2{}^{17}O$	3.692800×10^{-4}	45.005278
NH_3	$^{14}NH_3$	0.995872	17.026549
	$^{15}NH_3$	0.003661	18.023583
NO	$^{14}N^{16}O$	0.993974	29.997989
	$^{5}N^{16}O$	0.003654	30.995023
	$^{14}N^{18}O$	0.001993	32.002234

NO_2	$^{14}N^{16}O_2$	0.991616	45.992904
	$^{15}N^{16}O_2$	0.003646	46.989938
SO_2	$^{32}S^{16}O_2$	0.945678	63.961901
	$^{34}S^{16}O_2$	0.041950	65.957695
H_2S	$H_2{}^{32}S$	0.949884	33.987721
	$H_2{}^{34}S$	0.042137	35.983515
	$H_2{}^{33}S$	0.007498	34.987105
O_2	$^{16}O_2$	0.995262	31.98983
	$^{16}O^{18}O$	0.003991	33.994076
	$^{16}O^{17}O$	7.422350×10^{-4}	32.994045

Abundance estimates from HITRAN2016 database (Gordon et al. 2017).

Given the molecular configuration of CO_2, the position of a given isotope in one of the two O atoms does not change its molecular energy. However, some molecules have distinct molecular configurations that lead to isotopomers when isotope substitution occurs. For example, N_2O has two N atoms that occur in different positions within the molecule (Figure 2.4). The ^{14}N in N_2O can therefore be substituted with ^{15}N at either position, resulting in two isotopomers of N_2O: $^{15}N\alpha$-N_2O and $^{15}N\beta$-N_2O (Figure 2.4).

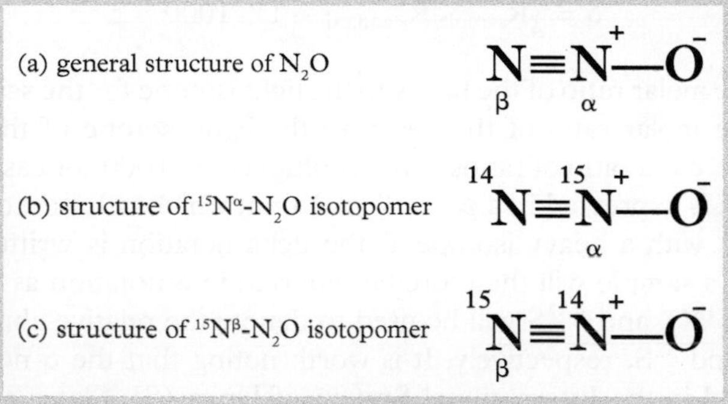

(a) general structure of N_2O $\overset{}{\underset{\beta}{N}}\equiv\overset{+}{\underset{\alpha}{N}}-\overset{-}{O}$

(b) structure of $^{15}N^{\alpha}$-N_2O isotopomer $\overset{14}{N}\equiv\overset{15\ +}{\underset{\alpha}{N}}-\overset{-}{O}$

(c) structure of $^{15}N^{\beta}$-N_2O isotopomer $\overset{15}{\underset{\beta}{N}}\equiv\overset{14\ +}{N}-\overset{-}{O}$

Figure 2.4 *Isotopomers of N_2O. ^{15}N can substitute for ^{14}N in both central (α) and terminal (β) positions of N in the N_2O molecule, resulting in $^{15}N\alpha$-N_2O and $^{15}N\beta$-N_2O.*

2.3 Why do we use delta notation?

The natural distribution of stable isotopes is such that one isotope is generally overwhelmingly more abundant than the other rare isotopes. Most often, the light isotopes are the most abundant while the heavy isotopes are rarer (Table 2.1). We are generally interested in the change in abundance of the heavy isotopes (H) relative to the light isotope (L). This is called the *isotopic ratio* (R):

$$R = H/L \tag{2.1}$$

Small differences in the isotopic composition of an environmental sample are difficult to report because of the very low relative abundance of the heavy isotope with respect to the light isotope. If reported as an R, the difference would often be at the level of the fourth or fifth decimal point. For example, the heavy hydrogen isotope, deuterium (^{2}H or D) in the Vienna Standard Mean Ocean Water Standard (VSMOW) is present at a concentration of 0.015574%, while the light isotope (^{1}H) is present at 99.984426% (Table 2.1). As water evaporates from the ocean into the atmosphere, more of the light isotope will become water vapor through a process called isotopic fractionation (see Chapter 3). Thus, the percentage of ^{2}H in water vapor may reduce even further. Can you imagine how hard it would be to report and discuss changes in the hydrogen isotopic ratio ($R = {}^2H/{}^1H$) value from 0.00015576 of the VSMOW to 0.00014953 of a hypothetical water vapor sample?

To address this challenge, the delta (δ) notation has become the most common notation used to report natural abundance isotopic values because it amplifies small differences in the relative abundance of the rare isotope. The δ notation was first introduced by McKinney et al. (1950) in order to report the measured isotopic values with reference to a standard (i.e., as a difference). The expression for the δ notation is:

$$\delta = (R_{sample/}R_{standard} - 1) \cdot 1000 \tag{2.2}$$

where R_{sample} is the molar ratio of the heavy to the light isotope for the sample of interest, and $R_{standard}$ is the molar ratio of the heavy to the light isotope of the standard (see Section 2.4). Thus, δ is a ratio of ratios and is multiplied by 1000 for easier reporting. In fact, the δ notation is expressed in a *per mill* (‰) unit, which is Latin for per thousand. For an element X, with a heavy isotope Y, the delta notation is written as $\delta^Y X$. The ^{13}C abundance of a sample will therefore be reported in δ notation as δ^{13}C. Similarly, δ^{15}N, δ^2H (δD), δ^{18}O, and δ^{32}S will be used to denote the relative abundance of ^{15}N, ^{2}H (or D), ^{18}O, and ^{32}S, respectively. It is worth noting that the δ notation does not use a unit accepted in the International System of Units (SI; Slater et al. 2001). For this reason, Brand and Coplen (2012) proposed the adoption of a new unit, called urey (symbol Ur), named after Harold C. Urey, to report δ values, rather than ‰. Since δ values are dimensionless, accordingly to SI δ is a "quantity of dimension 1" (Dybkaer 2003). The urey would thus be considered a quantity of 1, and a value of ‰ would be reported as a milliurey (mUr). However, this proposal has not found much acceptance in ecology and biogeochemistry, and natural abundance stable isotopes are still commonly reported using the δ notation in *per mill* (‰).

While δ values are standardized to make isotopic ratios easier to handle, they may not be intuitive at first. So, how does this notation work? Given this formula, it becomes clear that the standard, by definition, has a δ value of 0‰. Any sample which is *enriched*

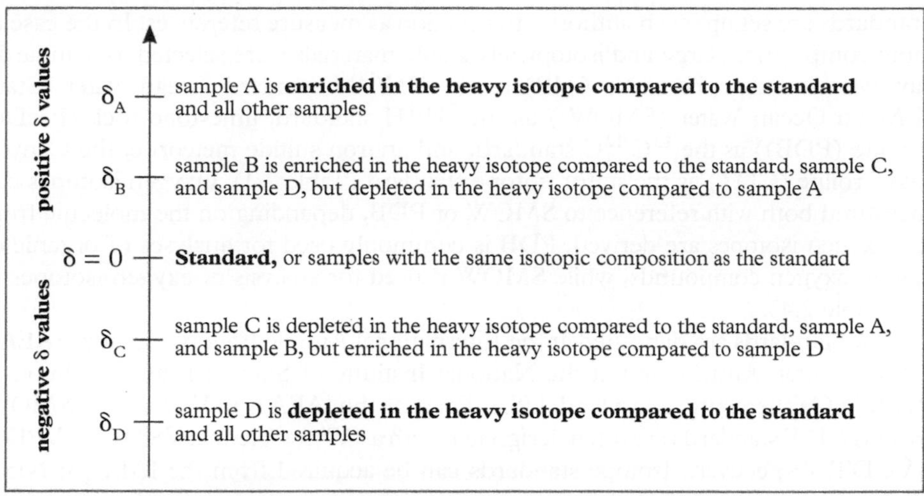

Figure 2.5 *The relationship between standards and samples with positive and negative δ values.*

in the heavy isotope as compared to the standard would have a positive δ value, while any sample which is *depleted* in the heavy isotope as compared to the standard would have a negative δ value. A useful thing to remember when familiarizing yourself with δ values is that *the lower the δ, the lighter the sample; the higher the δ, the heavier the sample* (Figure 2.5).

Let's go back to our hypothetical water molecule which evaporated from the VSMOW (Table 2.4) with an R of 0.00014953. Knowing the R of VSMOW, the standard used for hydrogen isotopes, we can calculate the water vapor δ^2H value as:

$$\delta^2\mathrm{H} = (0.00014953/0.00015576 - 1) \cdot 1000 = -40‰ \quad (2.3)$$

Clearly, a value of −40 is much easier to handle than 0.00014953. Additionally, just by the fact that it is negative (i.e., lower than the ocean water which we use as a standard that by definition has a δ value of 0‰), we can immediately say that the water vapor is 4% (40‰) lighter (i.e., depleted in the heavy isotope, ^{2}H), than the ocean water standard from which it evaporated (Figure 2.5).

2.4 Isotope standards

When first introduced, the δ notation was used to identify sample variations in isotopic abundance from a laboratory working standard (McKinney et al. 1950). It quickly became evident that the use of global standards would enable comparisons of δ values across different laboratories. To this end, committees of stable isotope geochemists worked to identify a set of internationally accepted standards beginning in the 1970s.

Standards are set up by an authority to function as measure references. In the case of isotopic composition, large and isotopically stable materials were selected as standards. Today, we use atmospheric N_2 (AIR) as the $^{15}N/^{14}N$ standard, ocean water (Standard Mean Ocean Water (SMOW)) as the $^{2}H/^{1}H$ standard, limestone rock (PeeDee Belemnite (PDB)) as the $^{13}C/^{12}C$ standard, and an iron sulfide meteorite, the Canyon Diablo Troilite (CDT) as the sulfur isotope standard (Table 2.4). Oxygen isotopes can be measured both with reference to SMOW or PDB, depending on the molecule from which oxygen isotopes are derived. PDB is commonly used for analyses of organic or inorganic oxygen compounds, while SMOW is used for analysis of oxygen isotopes in water (Table 2.4).

Isotope standards are deposited at the International Atomic Energy Agency (IAEA), based in Vienna, Austria, and at the National Institute of Standards and Technology (NIST), in Gaithersburg, Maryland, USA. Because the IAEA is in Vienna, the SMOW, PDB, and CDT standards are often designated with a "V" for Vienna: VSMOW, VPDB, and VCDT, respectively. Isotope standards can be acquired from the IAEA or NIST

Table 2.4 *Isotopic composition for international reference standards (H and L correspond to heavy and light isotopic components, respectively).*

International reference standard	Isotopic ratio (H/L)	Value (H/L)	H (%)	L (%)
Standard Mean Ocean Water (SMOW) and Vienna Standard Mean Ocean Water (VSMOW)	$^{2}H/^{1}H$	0.0001558	0.015574	99.984426
	$^{17}O/^{16}O$	0.0003799	0.03790	99.76206
	$^{18}O/^{16}O$	0.0020052	0.20004	99.76206
PeeDee Belemnite[a] (PDB) and Vienna PeeDee Belemnite (VPDB)	$^{13}C/^{12}C$	0.0112372	1.109915	98.890085
	$^{17}O/^{16}O$	0.0003859	0.0385	99.7553
	$^{18}O/^{16}O$	0.0020672	0.2062	99.7553
Atmospheric N_2 (AIR)	$^{15}N/^{14}N$	0.0036765	0.36630	99.63370
Canyon Diablo Troilite[b] (CDT) and Vienna Canyon Diablo Troilite (VCDT)	$^{33}S/^{32}S$	0.0078772	0.74865	95.03957
	$^{34}S/^{32}S$	0.0441626	4.19719	95.03957
	$^{36}S/^{32}S$	0.0001533	0.01459	95.03957

[a] Limestone from the PeeDee Formation in South Carolina (derived from the Cretaceous marine fossil Belemnitella americana);
[b] The Canyon Diablo Troilite, is an iron sulfide meteorite that impacted at Barringer Crater, Arizona. Fragments were collected around the crater and nearby Canyon Diablo.
Table adapted from Gröning (2004).

but are very expensive and available in very limited amounts. Thus, isotope laboratories commonly use working standards to calibrate their measurements.

2.5 When should we use fractional and atom percent notations?

Heavy isotope abundance can also be reported in proportion to the total element abundance (i.e., H+L), termed the *fractional notation*:

$$^{H}F = H/(H + L) \tag{2.4}$$

When the fraction of the heavy isotope is expressed in percentage, it is defined as the heavy atom % notation (^{H}AP):

$$^{H}AP = {}^{H}F \cdot 100 \tag{2.5}$$

Atom % reports stable isotope abundance as a molar fraction and therefore it is an accepted SI unit (Slater et al. 2001).

The atom % notation is linearly related to the δ notation at low heavy isotope abundances typical of the natural range of heavy isotopes (Figure 2.6). Thus, for natural isotope abundance values both atom % and δ units can be used. The δ notation is preferred for ease of use and standardization. δ values can be converted to atom % values as follows:

$$^{H}AP = 100 \cdot (\delta + 1000)/[(\delta + 1000) + (1000/R_{standard})] \tag{2.6}$$

The variation in natural abundance of isotopes differs between different biogeochemical pools due to a process called isotopic fractionation. We will learn in Chapter 3 how we can use this variation in natural abundance to understand biogeochemical and ecological processes. Heavy isotopes can also be artificially added to a sample, increasing the heavy isotope abundance to values outside that of the natural range (see Chapter 5). For samples highly enriched in the heavy isotope, the linearity between the δ and the atom % notation is lost, and the δ notation becomes inaccurate. Similarly, it is erroneous to use the δ notation to calculate isotopic discrimination (Chapter 3) and in mixing models (Chapter 4) with high heavy isotope enrichment values. In all these cases the atom % notation must be used. Additionally, because the δ notation is a ratio of ratios it cannot be used to calculate the coefficient of variation, in which case the atom % notation must be used. We refer the readers to Fry (2006) for a detailed description of the correspondence among isotope notations.

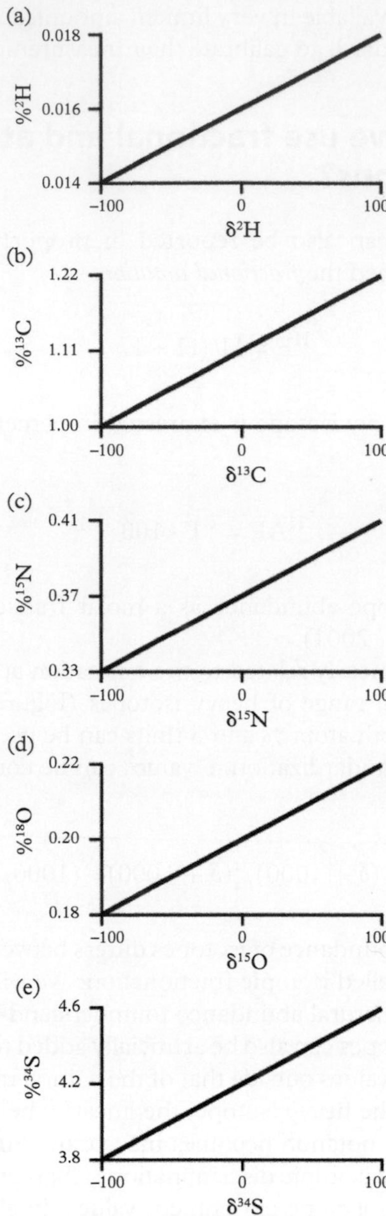

Figure 2.6 *Linear relationships between atom % and δ notation for ^{2}H (a), ^{13}C (b), ^{15}N (c), ^{18}O (d), and ^{34}S (e).*
Modified from Fry (2006).

References

Brand, W. A. and T. B. Coplen (2012). "Stable isotope deltas: tiny, yet robust signatures in nature." *Isotopes in Environmental and Health Studies* **48**(3): 393–409.

Dybkaer, R. (2003). "Units for quantities of dimension one." *Metrologia* **41**(1): 69–73.

Fry, B. (2006) *Stable isotope ecology*. New York: Springer.

Gordon, I. E., L. S. Rothman, C. Hill, R. V. Kochanov, Y. Tan, P. F. Bernath, M. Birk, V. Boudon, A. Campargue, K. V. Chance, B. J. Drouin, J. M. Flaud, R. R. Gamache, J. T. Hodges, D. Jacquemart, V. I. Perevalov, A. Perrin, K. P. Shine, M. A. H. Smith, J. Tennyson, G. C. Toon, H. Tran, V. G. Tyuterev, A. Barbe, A. G. Császár, V. M. Devi, T. Furtenbacher, J. J. Harrison, J. M. Hartmann, A. Jolly, T. J. Johnson, T. Karman, I. Kleiner, A. A. Kyuberis, J. Loos, O. M. Lyulin, S. T. Massie, S. N. Mikhailenko, N. Moazzen-Ahmadi, H. S. P. Müller, O. V. Naumenko, A. V. Nikitin, O. L. Polyansky, M. Rey, M. Rotger, S. W. Sharpe, K. Sung, E. Starikova, S. A. Tashkun, J. V. Auwera, G. Wagner, J. Wilzewski, P. Wcisło, S. Yu, and E. J. Zak (2017). "The HITRAN2016 molecular spectroscopic database." *Journal of Quantitative Spectroscopy and Radiative Transfer* **203**: 3–69.

Gröning, M. (2004). "International stable isotope reference materials." In: P. A. de Groot, ed. *Handbook of stable isotope analytical techniques*. Amsterdam, Elsevier: 874–906.

McKinney, C. R., J. M. McCrea, S. Epstein, H. A. Allen and H. C. Urey (1950). "Improvements in mass spectrometers for the measurement of small differences in isotope abundance ratios." *Review of Scientific Instruments* **21**: 724–730.

Slater, C., T. Preston and L. T. Weaver (2001). "Stable isotopes and the international system of units." *Rapid Communications in Mass Spectrometry* **15**(15): 1270–1273.

Soddy, F. (1923). *The origins of the conception of isotopes*. Stockholm, Les Prix Nobel en 1921–1922.

Activity: Using Isotope Notation in Ecology

Objectives

- Understand how to calculate and convert between δ and atom % isotope notations in ecological contexts.
- Interpret variation in isotopic differences among ecosystem components.

Tools and background

This activity can be completed entirely by hand with only a calculator. However, we encourage you to complete this activity using a spreadsheet or data processing software that will allow you to make better use of formulae and functions for converting between isotope notations. Developing these computational tools will improve your understanding of the relationships between isotope notations and standards. Once developed, you can save these tools to facilitate any stable isotope data analysis efforts you undertake in the future.

Exercises

Consider the following examples of C, N, O, and H isotopes in a variety of ecological systems (Figures A2.1–A2.4). Each example provides isotopic information in either δ or atom % notation. Using the isotopic values provided, the notation formulae (see Sections 2.3 and 2.5), and data on isotopic composition of international reference standards (see Section 2.4), calculate the isotopic value for the missing notation (i.e., convert from δ to atom % and vice versa). Use Table A2.1 as a guide to develop a spreadsheet or code to complete the conversions. The values given are realistic but were made up for the purpose of this activity.

Review questions

1. Review each ecological system (terrestrial C pools, terrestrial N pools, continental effects in the water cycle, and soil and plant water dynamics). Within each system, which component is the most enriched in the heavy stable isotope of interest? Which component is most depleted?
2. Review the stable isotope values across systems. Which element has the largest range of stable isotope values?
3. Looking at the isotope numbers in your spreadsheet, which notation provides isotope values that are easier to handle and interpret?

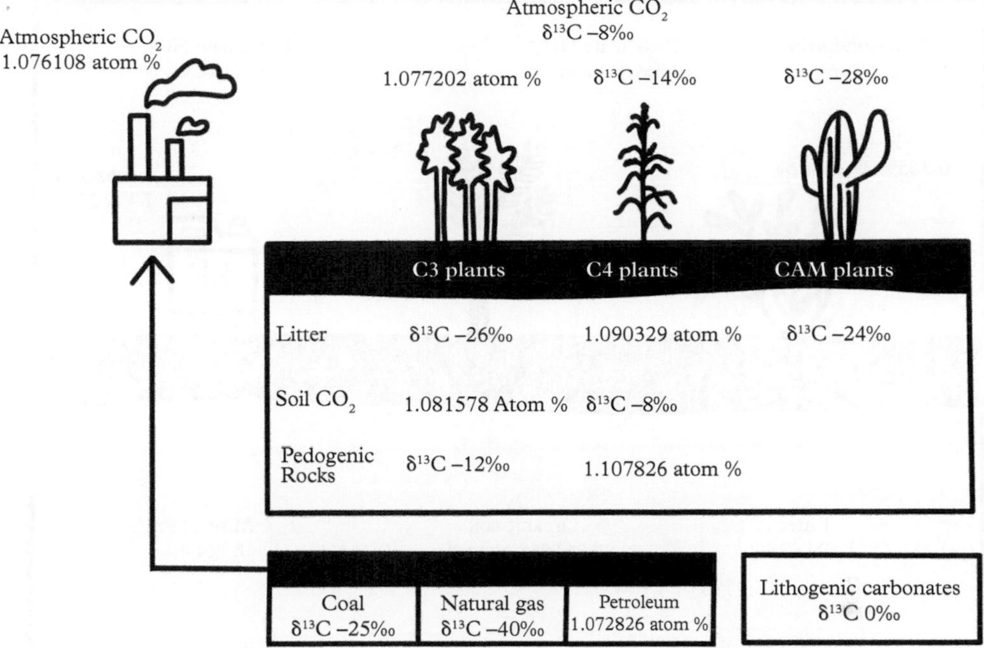

Figure A2.1 *^{13}C isotopic composition of various terrestrial C pools. Some isotope values are given in δ notation, while others are given in atom % notation. CAM, crassulacean acid metabolism.*

Table A2.1 *Example spreadsheet used to convert between δ and atom % isotope notations for selected terrestrial C pools. Given values are entered from conceptual Figures A2.1–A2.4, calculated values must be derived from appropriate formulae*

Ecosystem component	δ (‰, given)	Atom % (given)	H/L (calculated)	δ (calculated)	Atom % (calculated)
Atmosphere	−8 ‰	1.076108			
C3 soil CO_2		1.081578			
C4 soil CO_2	−8 ‰				
C3 plant		1.077202			
C4 plant	−14 ‰				
CAM plant	−28 ‰				
C3 plant litter	−26 ‰				
C4 plant litter		1.090329			
CAM litter	−24 ‰				

CAM, crassulacean acid metabolism.

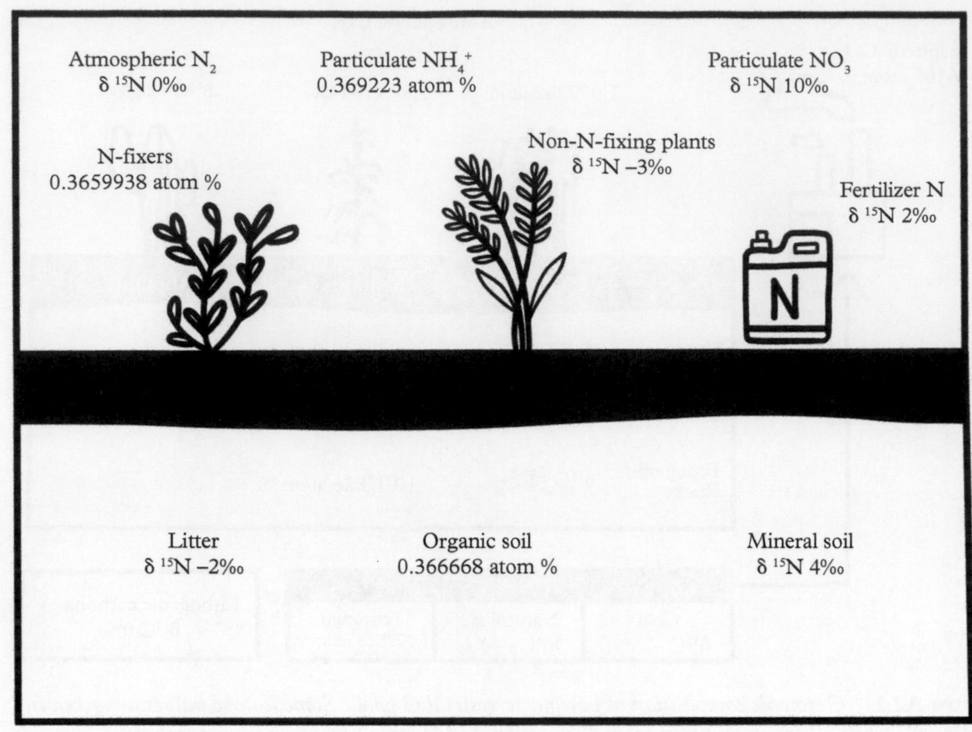

Figure A2.2 *^{15}N isotopic composition of various terrestrial N pools. Some isotope values are given in δ notation, while others are given in atom % notation.*

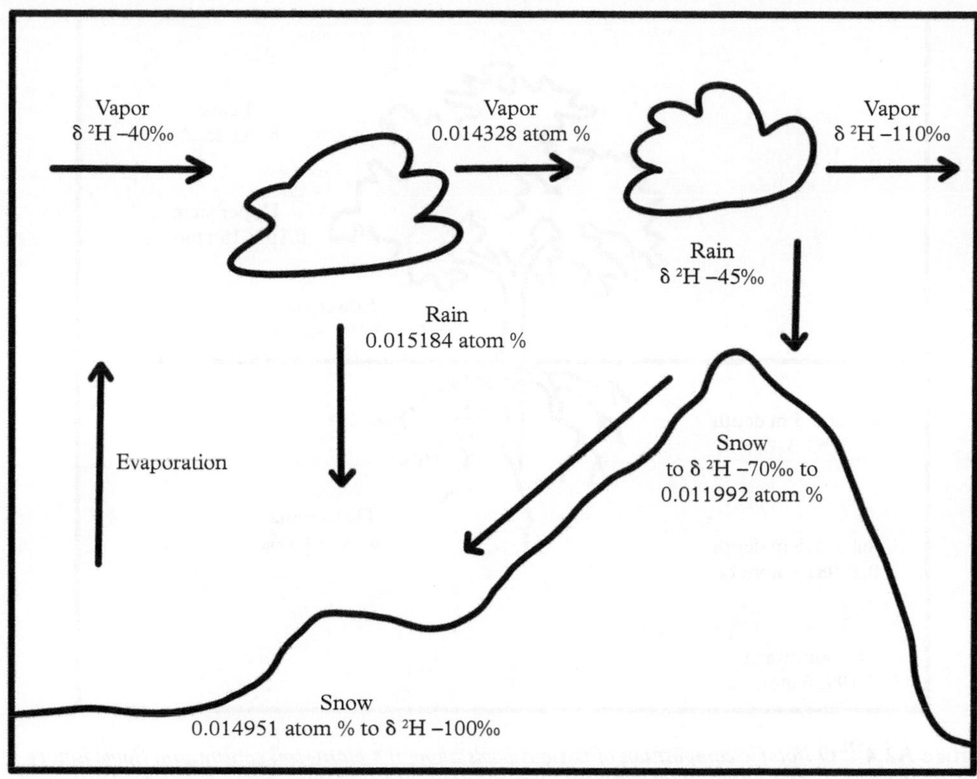

Figure A2.3 *^{2}H isotopic composition of various pools in the water cycle along a continental gradient. During condensation of cloud vapor to form rain, the heavy isotopes of H and O accumulate in the rain, while the remaining vapor becomes more depleted as they move inward to a continent. Some isotope values are given in δ notation, while others are given in atom % notation.*

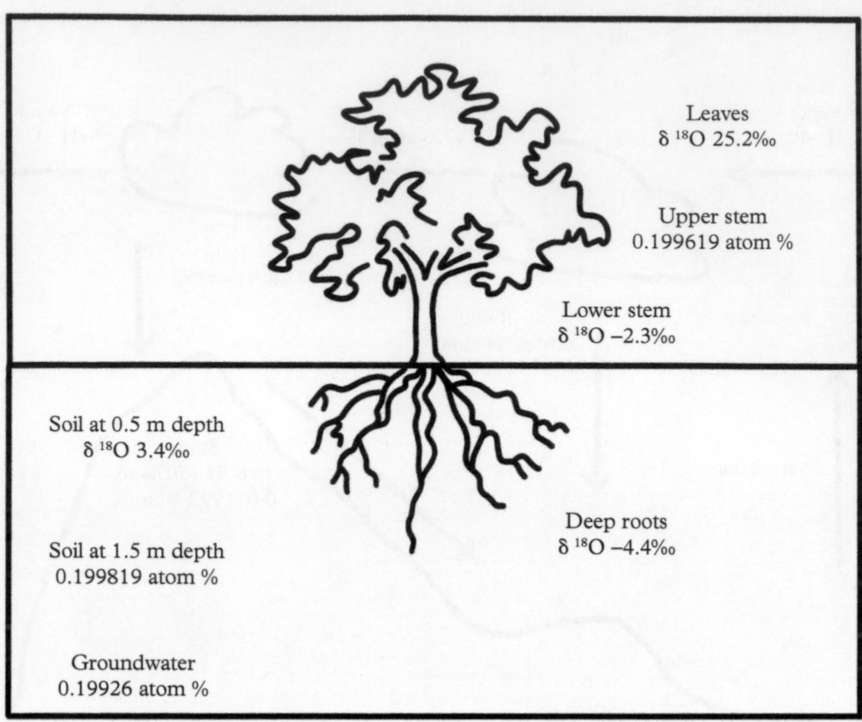

Figure A2.4 *^{18}O isotopic composition of various pools along the plant–soil continuum. Some isotope values are given in δ notation, while others are given in atom % notation.*

3

Isotopic Fractionation

3.1 Stable isotope fractionation

The varied isotopic composition in molecules has widespread consequences for how the elements that make up those molecules cycle through the environment. As we learned in Chapter 2, isotopocules of the same molecule have nearly identical behaviors in the environment (Section 2.2). However, the tiny differences in the mass and/or energy of isotopocules may affect their behavior just enough to result in *isotopic fractionation* during physical and chemical processes. Isotopic fractionation is the preferential reactivity of one isotope versus another, so that the reaction rate of one isotope is faster than the other. In this chapter, we focus on the atomic-level mass-dependent isotopic fractionation of isotopologues during chemical and physical processes. We discuss fractionation processes that form isotopomers, where the relative abundance of one isotopomer versus another varies, rather than the overall heavy isotope enrichment, due to site-specific (also known as non-random) distribution of heavy isotopes during molecular biosynthesis (e.g., N_2O production, glucosynthesis).

In rare cases, isotopic fractionation may occur independently of the isotopic mass differences. In nature, this mass-independent fractionation was first observed with respect to ^{17}O and ^{18}O isotope effects in ozone formation (Thiemens and Heidenreich 1983) and later for sulfur isotopes in Archean sediments (Pavlov and Kasting 2002), mainly with implications for atmospheric science and astrophysics, respectively. Mass-independent fractionation is not a common driver of natural abundance isotopic differences on Earth, and therefore we will not discuss it in this book.

Isotopic fractionation generates the variability in the natural abundance of isotopes in different compartments of their biogeochemical cycles. Fractionation is the reason we study and apply isotopes in ecology and biogeochemistry. As Brian Fry wrote in his landmark book, *Stable Isotope Ecology* (2006): "Fractionation creates the artist's palette, the isotope colors that are later mixed and arrayed to form the grand isotope masterpieces of nature."

The nature of isotopic fractionation in the environment is varied. Isotope fractionation may occur during physical processes or chemical reactions. It may occur in open or closed systems that influence the degree to which fractionation is realized in

A Primer on Stable Isotopes in Ecology. M. Francesca Cotrufo and Yamina Pressler, Oxford University Press.
 DOI: 10.1093/oso/9780198854494.003.0003

a given environment. Further, isotope effects can happen in non-equilibrium processes or in thermodynamic equilibrium. For these reasons, a clear understanding of isotopic fractionation is key to applying stable isotopes to answer ecological and biogeochemical research questions.

In this chapter, we will present the basis of isotope fractionation for the use of natural abundance isotope variation in ecological and biogeochemical studies. We also discuss how fractionation factors are calculated. We use C3 and C4 photosynthesis as case studies to illustrate isotopic fractionation of $^{13}CO_2$ versus $^{12}CO_2$. Isotope fractionation during photosynthesis demonstrates both physical and chemical fractionation processes, as well as instances of both open and closed systems. At the end of this chapter, we will discuss site-preferential fractionation of ^{15}N forming N_2O isotopomers during denitrification and the non-random distribution of ^{13}C in glucose isotopomers and how it determines differences in the ^{13}C natural abundance of organic compounds (e.g., lipids versus sugars).

3.2 Physical fractionation

In physics, motion describes the movement of a body, which is dependent on its mass. Thus, physical processes involving the motion of an atom or a molecule fractionate on the basis of its mass. Two isotopologues of the same molecule, for example, $^{13}CO_2$ and $^{12}CO_2$, will therefore move at different speeds or require a different amount of force to move. The heavy isotopologue will move slower or require more force.

In nature, molecules move from regions of high concentration to regions of low concentration through a process called diffusion. As described in Fick's first law of diffusion, diffusion is determined by the concentration gradient of the diffusing molecule and a coefficient of diffusion, which is inversely related to the mass of that molecule. Because of this mass dependency, different isotopologues will diffuse at different speeds depending on their isotopic composition. The molecule with the lighter isotope will diffuse faster and the one with the heavy isotope will diffuse slower. This generates isotope fractionation during diffusion proportional to the mass difference of the isotopes or isotopologues involved in the diffusion process. Therefore, gas molecules like CO_2 diffuse from the soil, where they are more concentrated, to the atmosphere, where their concentration is much lower. As CO_2 diffuses from the soil, the lighter isotopologues (e.g., $^{12}CO_2$) will diffuse faster, generating isotopic fractionation. When the system reaches steady state, the CO_2 in the soil air will be enriched in ^{13}C as compared to the CO_2 that diffuses into the atmosphere (Amundson et al. 1998).

The velocity of a body in motion is also dependent on the number of collisions that the body experiences during the motion. The more collisions, the less the mass of the body affects its motion. Thus, physical fractionation is expressed the most in a vacuum, less in air, and is not detectable in water. For example, isotope measurements by mass spectrometry use the difference in mass among isotopologues (e.g., of CO_2, N_2, O_2) to detect them as they are accelerated in a magnetic field. To maximize the different mass response to acceleration (i.e., to operate under maximum physical fractionation), isotope ratio mass spectrometers operate under vacuum, as we will learn in Chapter 6.

State changes also lead to isotopic fractionation. Because isotopologues containing the heavy isotopes are slower and less reactive than the lighter ones, they require more energy to change phase and tend to stay in the phase with the lowest free energy. Isotope fractionation during state change is commonly used in the study of the water cycle (Gat et al. 1996). However, while liquid to vapor water isotopic fractionation is relatively easy to measure, fractionation during solid to liquid transitions is harder to determine. A 2.1% and a 0.3% enrichment of ^{2}H and ^{18}O, respectively, has been found in solid (ice) as compared to liquid water (Wang and Meijer 2018). The Rayleigh distillation process is a typical state transition isotopic fractionation (Rayleigh 1902), which we will describe below as an example of fractionation in closed systems with respect to the substrate input.

State transitions may reach equilibrium. At equilibrium, the lighter isotopes will preferentially reside in the phase with the higher free energy, and fractionation will be higher the more pronounced the energetic difference between the two states is. Thus, the equilibrium fractionation will depend on the energy (i.e., temperature) of the system. The equilibrium fractionation factors for the ^{2}H and ^{18}O in liquid to gas state transitions of water at 20°C are reported in Figure 3.1. How would they change at higher temperatures? As the temperature increases, the free energy of liquid water rises and the energetic difference between the liquid and gas water diminishes, making it easier for the heavy water molecules to vaporize. Thus, equilibrium fractionation effects decrease with increasing temperature. This generates the possibility of using water isotopes as a historical thermometer! In fact, thanks to the different fractionation between ocean water and water vapor in hot and cold years, which results in the deposition of polar ice enriched in ^{18}O in hot years and depleted in ^{18}O in cold years, scientists have been able to reconstruct the temperature on Earth for the past hundreds of thousands of years (Andersen et al. 2004).

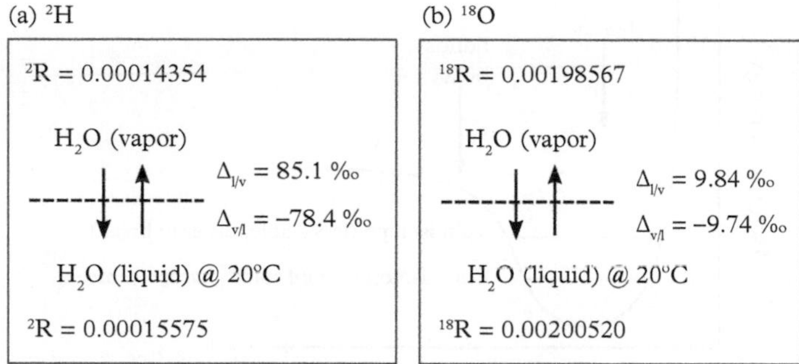

Figure 3.1 *Equilibrium fractionation factors for ^{2}H (a) and ^{18}O (b) in liquid to vapor and vice versa state transitions of water at 20° C.*
Modified from Mook and Rozanski (2000).

In nuclear chemistry, the isotope composition differences between gaseous and condensed phases at equilibrium are defined as vapor pressure isotope effects (Vértes et al. 2011). As said above, typically lighter isotopologues are more volatile than heavier ones. However, in some cases the reverse is true, and an "inverse" vapor pressure isotope effect is observed. Inverse isotope effects have, for example, been observed during pure-phase evaporation of halogenated organic compounds (Horst et al. 2016).

3.3 Chemical fractionation

Chemical fractionation occurs during the breaking and forming of bonds. Heavier isotopes form bonds with lower vibrational frequencies and higher binding energy. Therefore, more energy is required to break as well as to form a bond involving a heavy isotope. When the isotopic composition affects the reactivity of a molecule, the product of the reaction has a different abundance of the heavy isotope compared to the substrate. In chemical terms, two isotopologues or isotopomers involved in a chemical reaction may have different reaction rate constants for the same reaction. The *kinetic isotope effect* refers to the effect of the isotope substitution, in terms of mass or position, on the reaction rate constant. This effect is commonly described by imagining the bond between two atoms as a well (Figure 3.2). The well represents a stable state of lower potential energy when two atoms are bonded compared to when the atoms are apart. The energetic state of the system increases when bonds are broken (Figure 3.2). Heavy atoms establish stronger bonds which fall deeper into the well. Heavy isotope bonds require more energy to form but are also more energetically stable and therefore require more energy to break. Consequently, many reactions involving the breaking and forming of isotope bonds discriminate against the heavy isotope which then preferentially accumulates in the substrate.

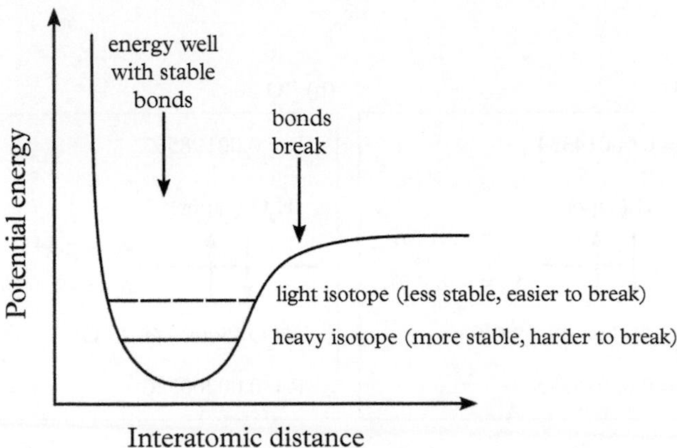

Figure 3.2 *The energy well analogy describing the kinetic isotope effect.*
Modified from Bigeleisen (1965).

Kinetic isotope fractionation was first observed in the electrolysis of water and is now amply studied to elucidate chemical reaction mechanisms. Later in this chapter, we will describe chemical fractionation during carboxylation reactions in photosynthesis. Other relevant examples in ecology and biogeochemistry are reactions involving N transformations, such as nitrification, denitrification, and assimilation when the lighter ^{14}N is preferentially transformed while ^{15}N remains in the substrate (Högberg 1997). Since these N transformations produce ^{15}N-depleted, mobile N molecules (e.g., NO_3^-, gaseous N) which can be easily lost from the system, kinetic N isotope effects lead to ^{15}N enrichment in the soil of ecosystems that experience high N losses (Craine et al. 2015).

Chemical reactions often happen in chains, where the product of one transformation is the substrate for the following reaction. Some chain processes are irreversible, in which case the kinetic fractionation of the most limiting step generally determines the kinetic fractionation of the entire process. For example, the first enzymatic reaction in the chain of reactions responsible for microbial toluene degradation was deemed to be the limiting step because it was found to be the major fractionating step (Morasch et al. 2001). In other cases, step reactions are reversible, and the overall isotopic fractionation effect depends on the fractionation of forward and backward fluxes at each chemical reaction step. The dissimilatory sulfate reduction pathway within sulfate-reducing microorganisms is an interesting example of reversible chain reactions resulting in the fractionation of sulfur isotopes (Rees 1973; Brunner and Bernasconi 2005).

An element's biogeochemical cycle often involves processes in which gases solubilize in liquids. For example, the solubilization of CO_2 in ocean water is a major process of the global C cycle. There are not many studies investigating isotope effects on the solubility of gases in liquids. However, in general, normal solubility isotope effects have been found, where the isotopologue containing the heavier isotope is more soluble than the corresponding lighter isotopologue. A notable exception is the solubility of CO_2 in water, where an inverse effect has been measured as the $^{12}CO_2$ solubilizes preferentially with respect to the $^{13}CO_2$ (Vogel et al. 1970).

Many chemical reactions are reversible and the products of one reaction are the substrate of the inverse reaction. At equilibrium, reversible chemical reactions accumulate the heavy isotope in the molecules with the highest bond strength. As explained above for physical equilibrium fractionation (Section 3.2), chemical equilibrium isotope effects are also highly dependent on the temperature of the system (Figure 3.3). Lower fractionation occurs as the temperature rises, and therefore the energy of the system increases. An example of the equilibrium isotope effect is the isotope fractionation in the chemical equilibrium reaction between CO_2 and bicarbonates (HCO_3^-; Eq. 3.1).

$$2CO_2 + 2H_2O \leftrightarrow 2H^+ + 2HCO_3^- \quad (3.1)$$

Bicarbonate is enriched in ^{13}C as compared to the CO_2 at equilibrium because it is the molecule with the strongest bonds (Figure 3.3). Because of this temperature-dependent equilibrium isotope effect, bicarbonates have a $\delta^{13}C$ approximately 8‰ higher (8.46‰ at 20°C) than the CO_2 they are in equilibrium with.

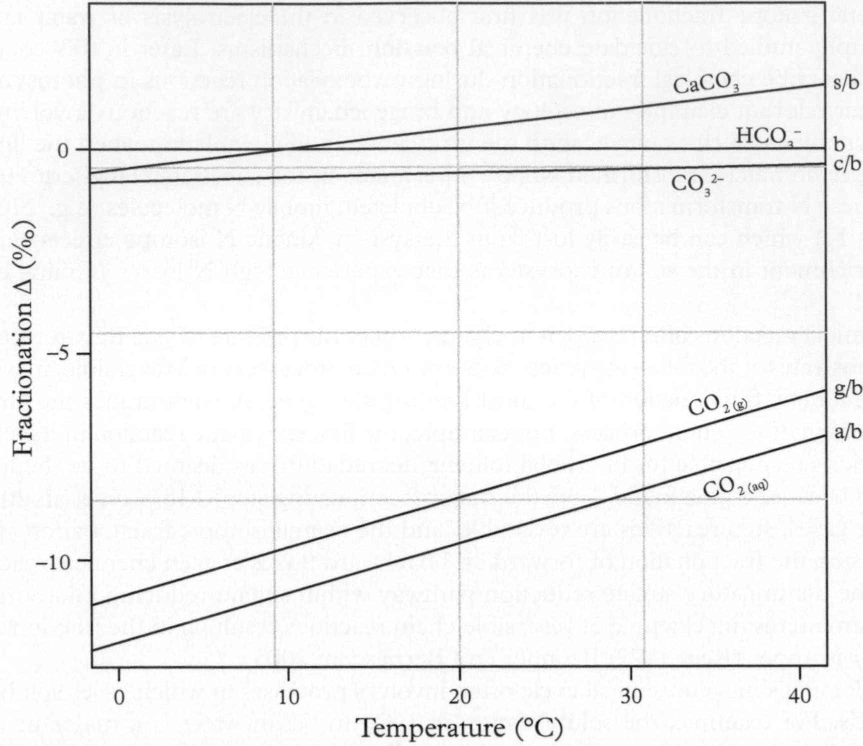

Figure 3.3 *Temperature-dependent equilibrium isotope fractionation for carbon isotopes. Lines show temperature-dependent fractionation for dissolved CO_2 (a), gaseous CO_2 (g), dissolved carbonate ions (c), and solid carbonate (s), all with respect to dissolved HCO_3^- (b).*
Modified from Mook and Rozanski (2000).

3.4 Fractionation factors

Quantification of isotope effects during physical and chemical reactions is typically pursued by nuclear physicochemists. Examples of approaches can be found in Vértes et al. (2011).

Ecologists and biogeochemists typically calculate isotope effects during either physical processes or chemical reactions, by comparing the isotopic ratios of the substrate (molecule A) to that of the product (molecule B) in the physical process or chemical reaction of interest. This is typically also applied to equilibrium reactions, as long as it is clear which is considered molecule A and which is molecule B at equilibrium. The isotope fractionation factor (α) is calculated as:

$$\alpha = {}^{A}R/{}^{B}R \quad (3.2)$$

where ^{A}R is the isotope ratio of the substrate and ^{B}R is the isotope ratio of the product. Therefore, when $\alpha > 1$ the heavy isotope is discriminated against during a physical process or chemical reaction. When $\alpha < 1$ the light isotope is discriminated against. In the absence of isotopic fractionation, $\alpha = 1$.

Because isotopic fractionation effects are typically small, fractionation factors are often close to 1. Therefore, the heavy isotope discrimination, or "big delta" (Δ), is most commonly used in ecology and biogeochemistry. It is calculated as:

$$\Delta = (\alpha - 1) \cdot 1000 \tag{3.3}$$

As for the δ notation, the Δ is reported in per mill (‰). A positive Δ signifies heavy isotope depletion during a physical process or chemical reaction, while a negative Δ indicates heavy isotope enrichment. For a "quick and dirty" estimate of the heavy isotope discrimination, Δ can be calculated as:

$$\Delta = \delta \mathrm{A} - \delta \mathrm{B} \tag{3.4}$$

However, a small error is introduced in this expression, and the accurate expressions to calculate the fractionation factor (α), or the heavy isotope discrimination (Δ) using the δ notation are, respectively:

$$\alpha = [(\delta A - \delta B)/1000] + 1 \tag{3.5}$$

$$\Delta = (\delta A - \delta B)/1 + (\delta B/1000) \tag{3.6}$$

In Section 3.6 below, we provide examples of the calculation of the isotope fractionation factor and heavy isotope discrimination for both physical and chemical processes during photosynthesis in C3 and C4 plants.

3.5 Fractionation in closed versus open systems

The extent to which isotope fractionation is expressed depends on the proportion of the substrate which undergoes a chemical or physical process. If all of the substrate reacts, fractionation is not expressed, and the product (molecule B) will have the same isotopic value as the substrate (molecule A). This implies two things:

1. Isotopic fractionation is constantly expressed when there is a continuous influx of the substrate (i.e., the system is open with respect to the substrate).
2. Isotopic fractionation between the substrate and the accumulated product diminishes with the progressive increase of the fraction of the substrate reacted until $\alpha = 1$ or $\Delta = 0$ when all the substrate has transformed into an accumulated product (i.e., the system is closed).

Let's provide some examples to make these concepts clearer. The efflux of CO_2 from the soil into the atmosphere is one of the most important C fluxes of the global biogeochemical C cycle. This process is estimated to transfer about 60 Pg C to the atmosphere globally every year (Schlesinger and Bernhardt 2013). The actual CO_2 efflux from the soil occurs through diffusion, while CO_2 is continuously produced in the soil by the respiration of the soil biota and plant roots, and the weathering of carbonates in calcareous soils. It is easy to recognize an open system where CO_2 (molecule A) is continuously produced to diffuse out of the soil into atmospheric CO_2 (molecule B). At steady state the diffusion rate will equal the rate of production of CO_2 within the soil (Figure 3.4). As we learned above, diffusion discriminates against the heavy molecule, proportional to the ratio of the light and the heavy masses, which for $^{12}CO_2/^{13}CO_2$ is 44/45. In air, the fractionation during diffusion of CO_2 is 4.4‰, meaning that at steady state the CO_2 flux into the atmosphere is 4.4‰ lighter than the soil CO_2 (Amundson et al. 1998; Figure 3.4).

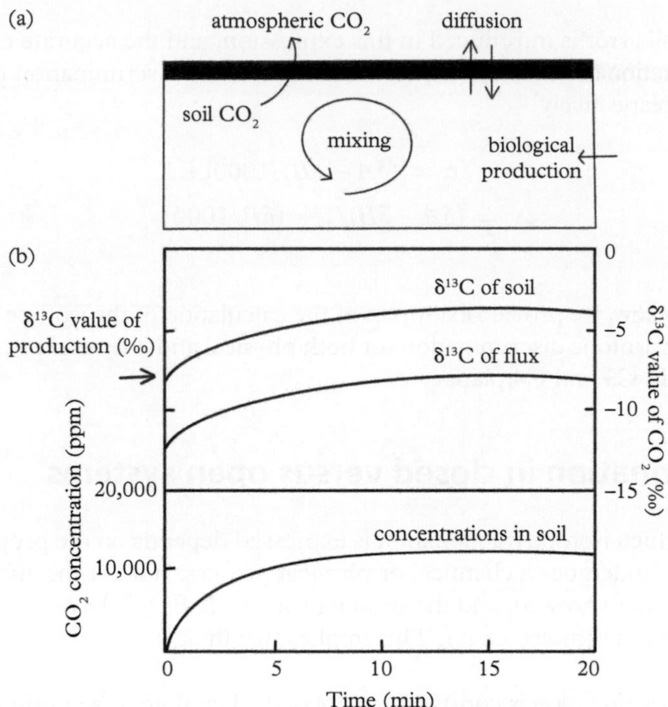

Figure 3.4 *Diagram illustrating a simple model for an open system: soil CO_2 efflux. In well-mixed soils, the $\delta^{13}C$ of the soil CO_2 efflux to the atmosphere equals that of biological production in the soil when the system reaches steady state (a). Sketch of a theoretical model illustrating isotope dynamics in soil where biological CO_2 production proceeds from no production (t = 0 min) to steady state (t = 20 min) (b).*

Modified from Amundson et al. (1998).

In nature, processes most commonly occur in open systems where the substrate is continuously replenished, as in the case of soil CO_2 efflux described above. However, there are cases in which a chemical or physical process happens in an environment where the amount of substrate decreases as it reacts, while the product is removed from the system (i.e., the system is open for the product but not for the substrate). In these cases, fractionation can be described by the Rayleigh distillation equation (Eq. 3.7). Lord Rayleigh described the ^{18}O isotopic fractionation during the distillation of a finite volume of liquid water with the vapor escaping the system. He proposed that at any time (t) during the distillation reaction, the isotopic ratio (R_t) of oxygen in the residual liquid water was a function of the fraction (f) of liquid water remaining, according to the equation:

$$R_t = R_0 f^{(1 - \alpha)} \tag{3.7}$$

where R_0 is the oxygen isotopic ratio of the initial water and α is the isotopic fractionation factor of water distillation at a given temperature. The Rayleigh distillation thus describes the progressive enrichment in ^{18}O of the residual liquid water sample as the lighter water molecules distill into vapor and escape. As a consequence, the instantaneous vapor product will enrich in ^{18}O as the fraction of residual water approaches 0 (Figure 3.5). Different forms of the Rayleigh distillation equation have been used to

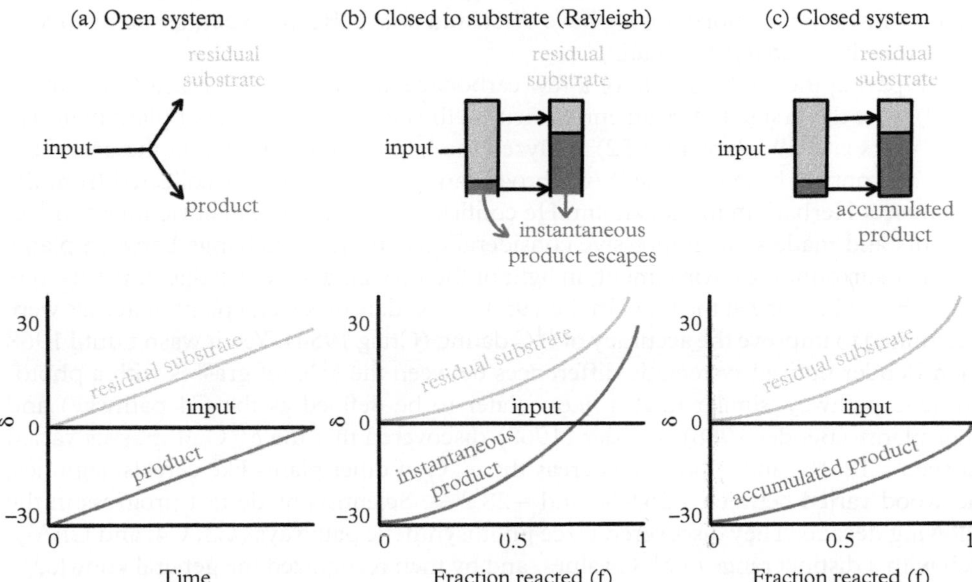

Figure 3.5 *Comparison of isotope dynamics in open (a), closed to substrate (b), and closed to substrate and product (c) systems. Dynamics in an open system are shown over time to steady state (a). Systems that are closed to the substrate (b) follow Rayleigh dynamics where the instantaneous product escapes. Delta values shown are arbitrary and included as an example to demonstrate the concept.*
Modified from Fry (2006).

estimate isotope effects and we refer readers to Scott et al. (2004) for an assessment of their performance.

In a completely closed system, where the product is not removed, the accumulated product will also progressively enrich. When all the substrate has reacted, the product will have the same isotopic value of the original substrate. *When all the substrate reacts into an accumulated product, fractionation will not be expressed.* Examples of closed systems are the ones that we create when we need to measure the isotopic value of a sample. As we will learn in Chapter 6, isotope ratios can only be measured on certain molecules. So, if we want to measure the $\delta^{13}C$ of a compound, let's say a plant leaf, we need to convert it to CO_2 through oxidation. In order to make sure that the isotope ratio of the product (CO_2) is exactly the same as that of the substrate (plant leaf) we need to make absolutely sure that fractionation is not expressed and thus we have to operate in a completely closed system until the reaction is complete (i.e., all the leaf C is oxidized to CO_2). Typically, this is done by coupling an elemental analyzer to an isotope ratio mass spectrometer (Section 6.3).

3.6 Isotopic fractionation at work: photosynthesis

The differential behavior of ^{12}C and ^{13}C during photosynthesis in C3 and C4 plants offers a complete and, therefore, often-highlighted example to explain all the different aspects of isotope fractionation we have discussed above. Reconstructing the history of how it was discovered is fascinating!

Comparing the $^{12}C/^{13}C$ ratio of a few carbonaceous samples, Nier and Gulbransen (1939) were the first to notice an enrichment in the light isotope (^{12}C) of plant material. A few years later, Wickman (1952) analyzed the $^{12}C/^{13}C$ ratio (note that the use of the δ unit had not yet been proposed) of a broad range of plant samples collected from the Riksmuseet Herbarium in Stockholm. He confirmed the widespread enrichment in ^{12}C in plants and made some impressive considerations on the C exchange between plants and their surrounding environment, in light of the very limited knowledge of plant–soil–atmosphere C cycling at the time. In the 1950s, $\delta^{13}C$ data on several plant materials were determined to improve the accuracy of ^{14}C dating (Craig 1954). Yet, it wasn't until 1968 when Bender noticed systematic differences between the $\delta^{13}C$ of grasses with a photosynthetic pathway 'similar to *Zea mays*' (later to be defined as the C4 pathway) and other plants (Bender 1968). Bender (1968) discovered that the $\delta^{13}C$ of grasses varied between $-12.2‰$ and $-14.3‰$ whereas the $\delta^{13}C$ of other plants like cereals, legumes, and wood varied between $-26.9‰$ and $-28.2‰$. Scientists made fast progress in the following decade. They discovered three photosynthetic pathways (C3, C4, and CAM), each with a distinct range of $\delta^{13}C$ values, and by then recognized the general knowledge of isotopic discrimination during enzymatic reactions. These efforts culminated with Graham D. Farquhar and colleagues' theory on C isotope discrimination during photosynthesis of C3 (Farquhar et al. 1982) and C4 (Farquhar 1983) plants—one of the most elegant theories ever presented in isotope biogeochemistry.

The isotopic discrimination of ^{13}C during C3 photosynthesis is described as:

$$\Delta = a + (b - a)\, p_i/p_a \tag{3.8}$$

where a is the fractionation occurring during CO_2 diffusion through the stomata, b is the net fractionation due to carboxylation and CO_2 solubilization, and p_a and p_i are the partial pressure of ambient and intercellular CO_2, respectively (Farquhar et al. 1982).

To illustrate this equation let's follow the journey of a C atom which begins with CO_2 diffusing through the boundary layer and stomata into the leaf (Figure 3.6). As we learned above, lighter isotopologues diffuse faster than their heavy counterparts, and in the case of CO_2 diffusion have an estimated Δ of 4.4‰. Thus a = 4.4‰ in Eq. 3.8 above. As CO_2 diffuses through the stomata inside the leaf, it will be 4.4‰ lighter than that which remains outside the leaf in the atmosphere (Figure 3.6). This results in seasonal variation in the $\delta^{13}C$ of atmospheric CO_2 more prominently in the Northern Hemisphere, where it enriches during the summer (i.e., plant growing season, when photosynthetic CO_2 uptake exceeds CO_2 efflux via respiration), and depletes during the winter season when respiration exceeds photosynthesis (Mook et al. 1983). Diffusion of CO_2 in water is not dependent on its mass (Section 3.2). Therefore, C isotope fractionation during diffusion in photosynthesis of aquatic plants is considered to be close to zero (O'Leary 1993).

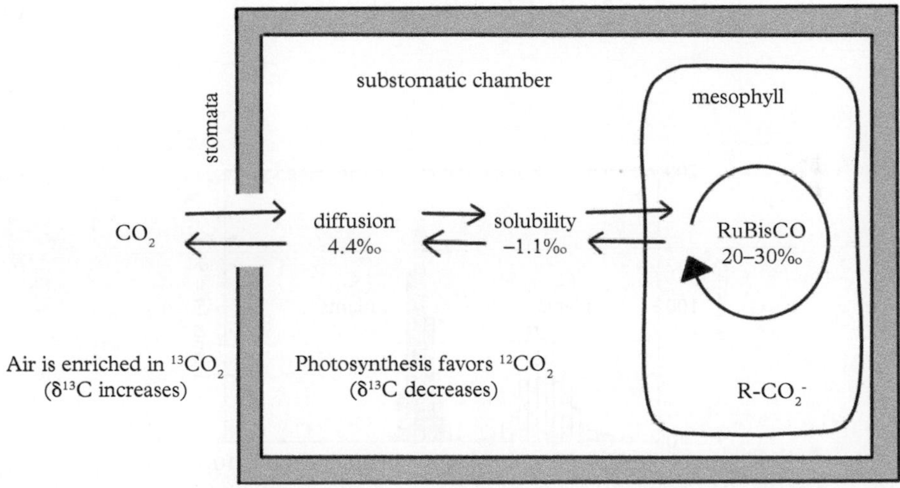

Figure 3.6 *Schematic of isotopic fractionation during C3 photosynthesis. The range of fractionation factors for the ribulose 1,5-bisphosphate carboxylase/oxygenase (RuBisCO) is reported. See text for additional detail on fractionation processes during C3 photosynthesis.*

Once in the plant, CO_2 dissolves in water. As reported in Section 3.3, this is a rare case of an inverse isotope effect where CO_2 dissolved in water is 1.1‰ lighter than CO_2 in air ($\Delta = -1.1‰$). The dissolved CO_2 then enters the mesophyll where it is fixed by the ribulose 1,5-bisphosphate carboxylase/oxygenase (RuBisCO) into 3-phosphoglycerate (3PGA), during the first step of C fixation of C3 plants. RuBisCO has a strong preference for $^{12}CO_2$, with reported discrimination values ranging from 20‰ to 30‰ (Farquhar et al. 1989). In Eq. 3.8, b represents both the fractionation during solubilization and carboxylation, and Farquhar et al. (1989) report a best fit of 27‰.

The most interesting part of Eq. 3.8 is that it links the isotopic fractionation with the ratio of CO_2 partial pressures inside and outside the leaf with a linear relationship. This means that when stomata are fully open, the ratio of p_i/p_a increases, the leaf operates as an open system, and therefore fractionation is fully expressed. In contrast, when the stomata close, the ratio of p_i/p_a decreases, the leaf operates as a closed system with respect to the substrate (CO_2), and fractionation between the CO_2 and the accumulated product (photosynthetate) decreases. This phenomenon generates the variability in $\delta^{13}C$ values of C3 plants despite the relatively constrained $\delta^{13}C$ of the atmospheric CO_2 (Figure 3.7). This also allows ecologists to use ^{13}C isotopic fractionation as a measure of water use efficiency in C3 plants (Dawson et al. 2002).

The isotopic fractionation of ^{13}C during C4 photosynthesis is different from that during C3 photosynthesis because of the different photosynthetic apparatus in these plants. It is described as:

$$\Delta = a + (c + b\varphi - a)\ p_i/p_a \tag{3.9}$$

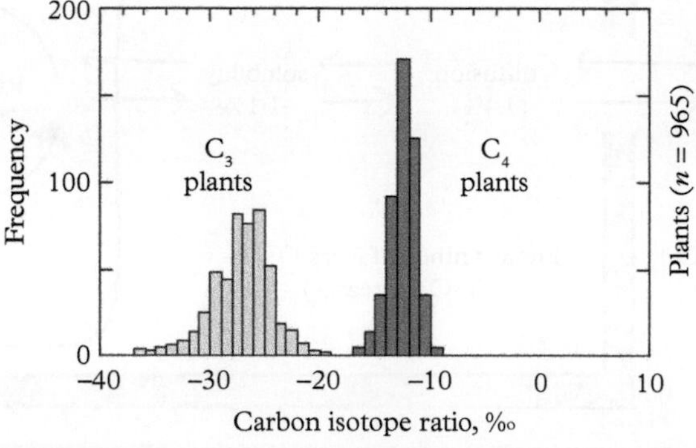

Figure 3.7 *Comparison of $\delta^{13}C$ values for C3 (light gray) and C4 plants (dark gray) for 965 plant species.*

Figure reproduced with permission from Cerling et al. (1997) and Ehleringer and Cerling (2002).

where a is the fractionation occurring during CO_2 diffusion through the stomata (as in C3 photosynthesis), c is the net fractionation due to C fixation by phosphoenolpyruvate (PEP) carboxylases of carbonic acid, b is the fractionation due to carboxylation by RuBisCO, φ is the bundle sheath cells leak factor, and p_a and p_i are the partial pressure of ambient and intercellular CO_2 (as in C3 photosynthesis), respectively (Farquhar 1983).

Let's now follow the journey of a C atom as it is fixed during photosynthesis of C4 plants to illustrate Eq. 3.9 (Figure 3.8). As for C3 plants, CO_2 diffuses through the stomata of C4 plants, with a diffusion fractionation (a) of 4.4‰. It dissolves and is converted to carbonic acid (HCO_3^-), with an equilibrium fractionation at 25°C of −7.9‰, which is then fixed by PEP carboxylases into oxalacetate. PEP carboxylase discriminates much less than RuBisCO, at around 2.2‰. The fractionation factor c accounts for both the above processes (conversion to carbonic acid and fixation by PEP carboxylase), resulting in a value of −5.7‰ (O'Leary et al. 1992). The oxalacetate is then transported into the bundle sheath cells, where it is reconverted to CO_2, which is finally fixed by RuBisCO.

Because the bundle sheath cells were thought to be impermeable to CO_2/HCO_3^-, it was originally assumed that RuBisCO carboxylation in bundle sheath cells, operating as

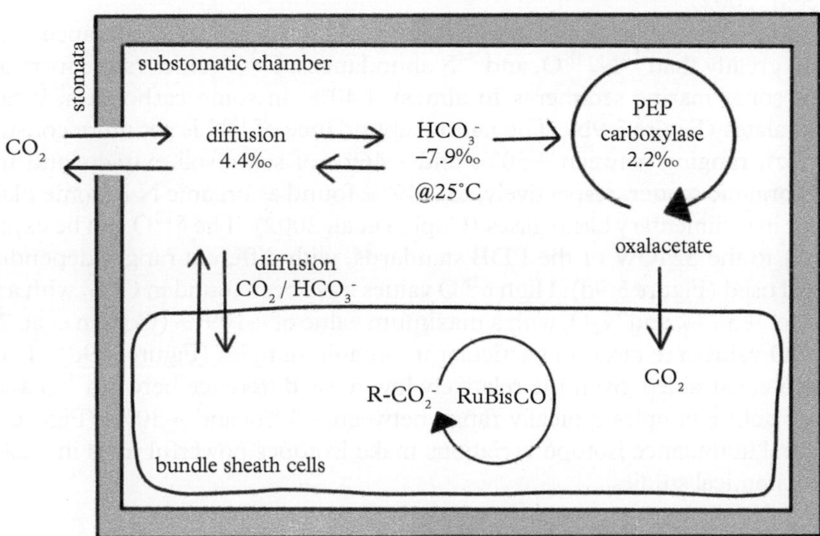

Figure 3.8 *Schematic of isotopic fractionation during C4 photosynthesis. The fractionation factor of phosphoenolpyruvate (PEP) carboxylase is reported. See text for additional detail on fractionation processes during C4 photosynthesis.*

a closed system, would not discriminate. However, measured $\delta^{13}C$ values of C4 plants were not justifiable by only the a and c fractionation factors. This led Farquhar and colleagues to recognize that some CO_2/HCO_3^- would leak out of the cells. As a result, the bundle sheath cells are considered an open system that requires the expression of RuBisCO fractionation (b) to be modified by the leakage factor (φ) (Farquhar 1983). Variation in stomata opening (i.e., p_i/p_a) and the bundle sheath leakage generates the observed variation in $\delta^{13}C$ of C4 plants, which is more constrained and always lower than $\delta^{13}C$ of C3 plants (Figure 3.7).

CAM plants can utilize either the C3 or C4 cycles; thus, their $\delta^{13}C$ values cover the range of both C4 and C3 plants (–34 ‰ to –10 ‰; Coplen et al. 2002).

3.7 Variation in the natural abundance of ^{2}H, ^{13}C, ^{15}N, ^{18}O, and ^{34}S isotopes

Because isotopes fractionate due to differences in their mass, the extent of fractionation and therefore their range of natural abundance variation depend on the mass difference between isotopes. As a consequence, natural abundance variation is largest for the lighter elements, and for the elements with the largest mass variation between their isotopes (Schauble 2004). It is therefore not surprising that in nature, hydrogen has the widest range of natural abundance isotope variation (Figure 3.9a). The $\delta^{2}H$ in nature can vary from as low as less than −800‰ in naturally occurring hydrogen gas to as high as +100‰ or +200‰ in some naturally occurring water and gases (Coplen et al. 2002).

The ^{13}C abundance has more constrained variation than ^{2}H abundance, but still varies more greatly than ^{15}N, ^{18}O, and ^{34}S abundances. $\delta^{13}C$ values range from almost −130‰ in some marine sediments to almost +40‰ in some carbonates, bicarbonates, and oxalates (Figure 3.9b). The natural abundance of ^{15}N is yet more constrained (Figure 3.9c), ranging between +50‰ and −46‰ of some soil extracts and marine particulate organic matter, respectively, to −49‰ found as organic N in some plants or animals and in sedimentary basin gases (Coplen et al. 2002). The $\delta^{18}O$ can be expressed with respect to the SMOW or the PDB standards, with different ranges depending on the standard used (Figure 3.9d). High $\delta^{18}O$ values have been found in CO_2, with a maximum value of +53‰, and N_2O, with a maximum value of +109‰ (Coplen et al. 2002). Negative ^{18}O values are rarer, in particular in organic samples (Figure 3.9d). The $\delta^{34}S$ range is also constrained given the relatively low mass difference between ^{32}S and ^{34}S, and organic sulfur samples typically range between −30‰ and +30‰ (Figure 3.9e). These natural abundance isotope variations make isotopes powerful tools in ecological and biogeochemical studies.

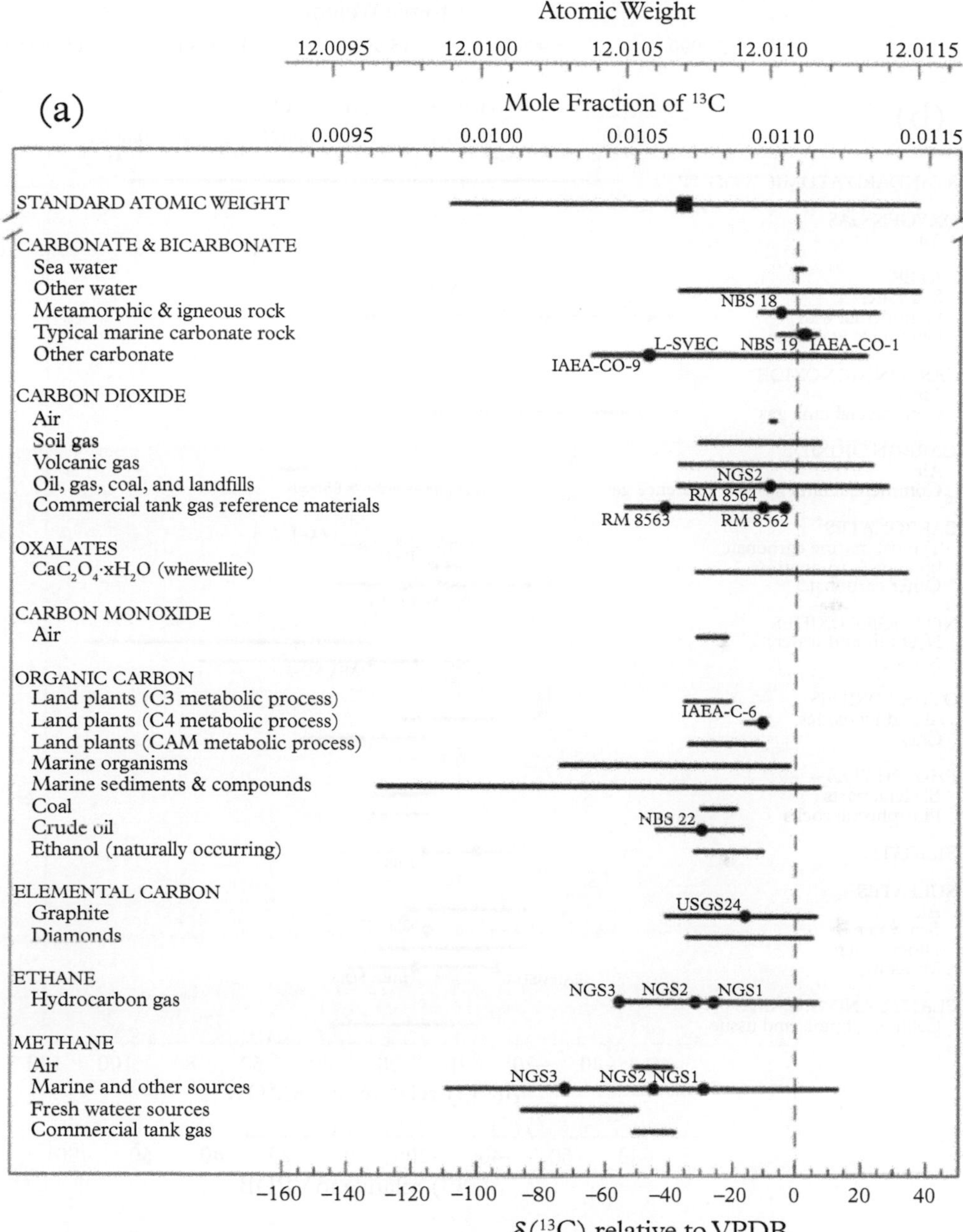

Figure 3.9 *Natural abundance variations for elements of interest: ^{13}C (a), ^{18}O (b), ^{15}N (c), ^{2}H (d), and ^{34}S (e).*

Reproduced with permission from Coplen et al. (2002).

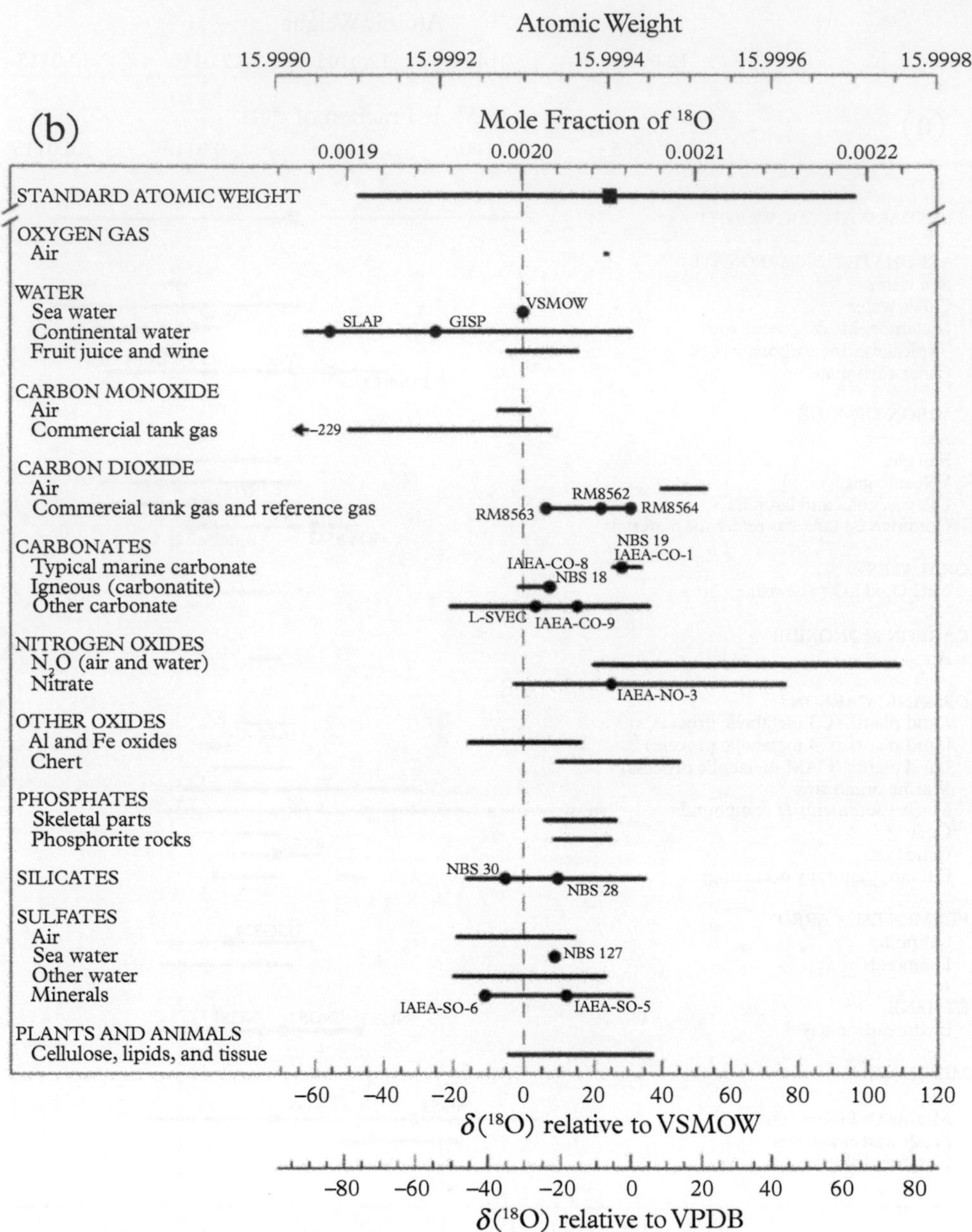

Figure 3.9 *Continued*

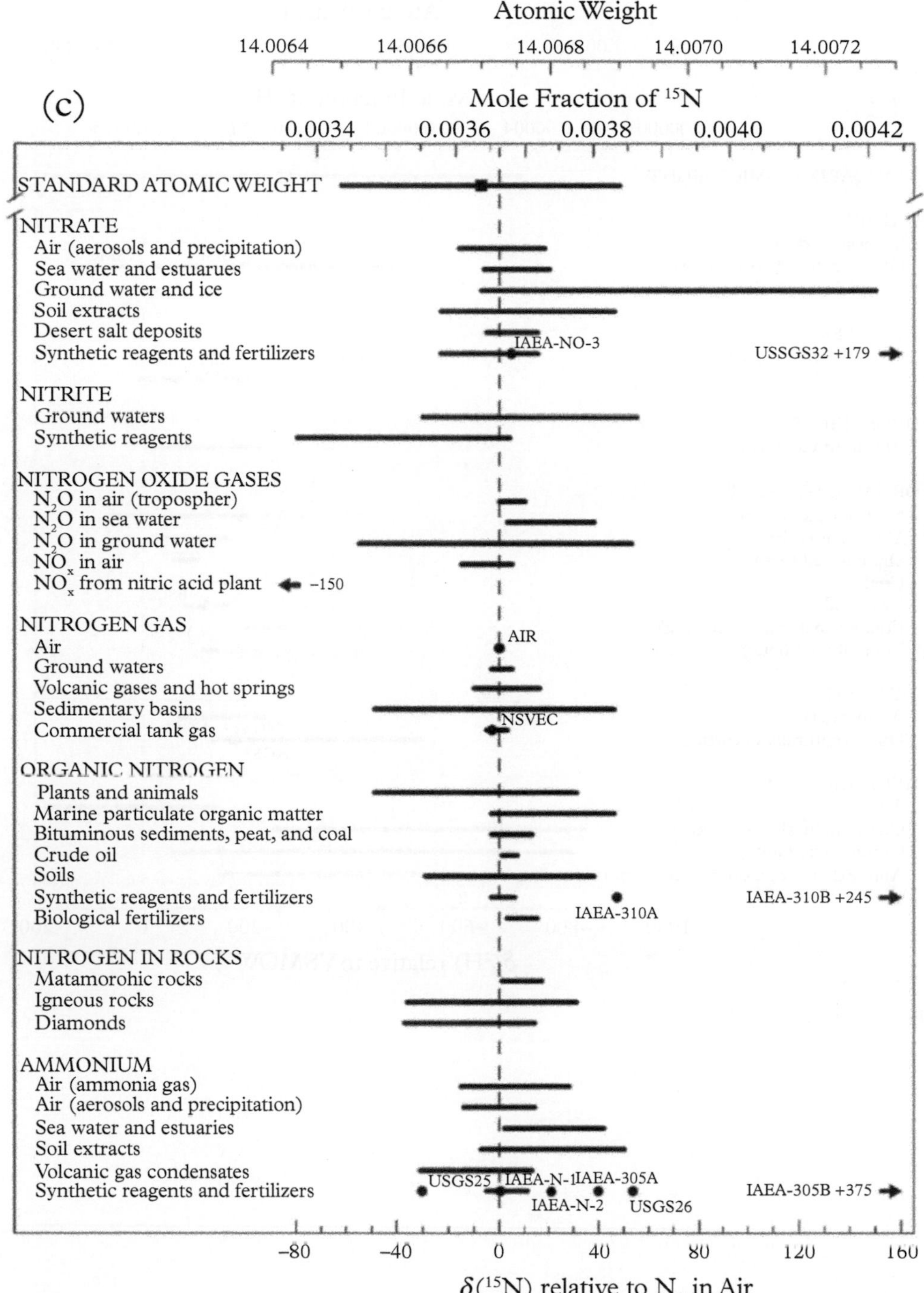

Figure 3.9 *Continued*

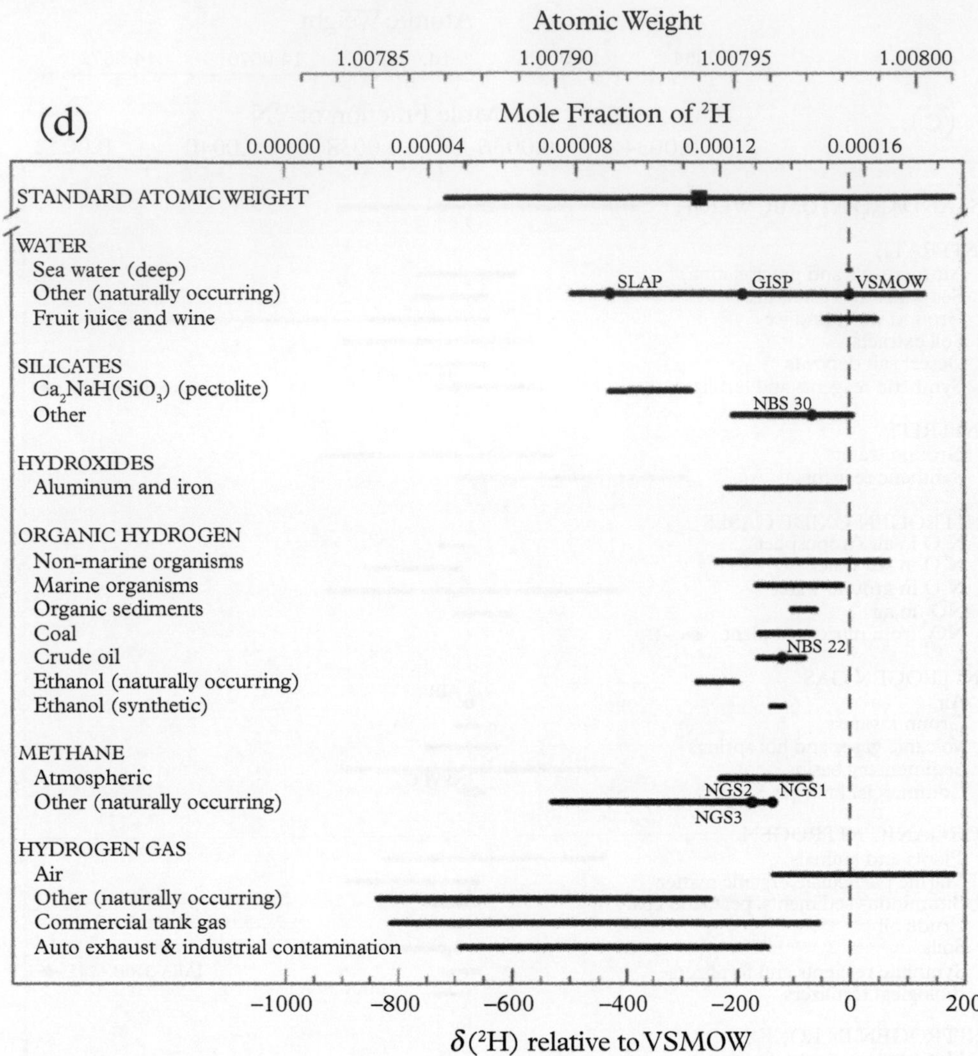

Figure 3.9 *Continued*

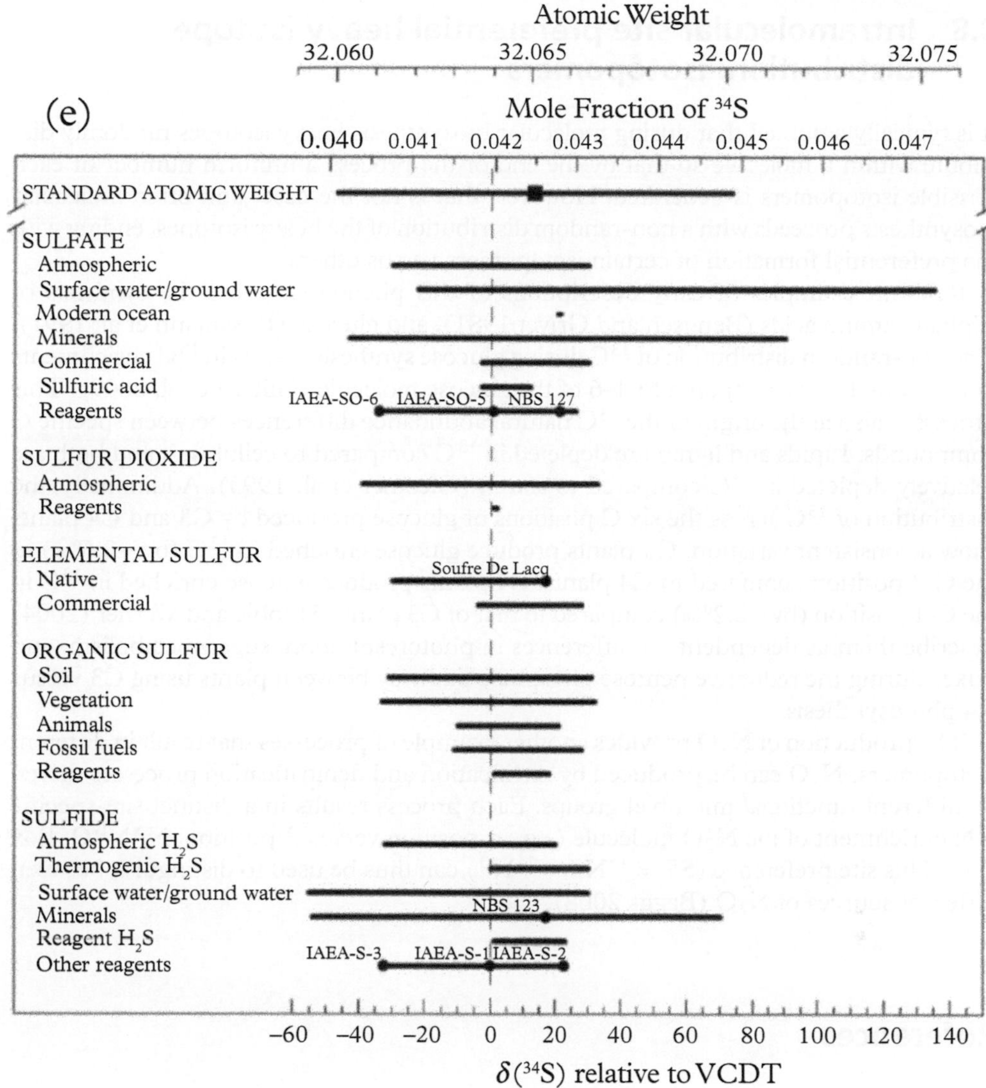

Figure 3.9 *Continued*

3.8 Intramolecular site preferential heavy isotope distribution: isotopomers

It is typically assumed that during molecular biosynthesis heavy isotopes randomly distribute within a molecule so that by the end of the process a uniform number of each possible isotopomers is generated. However, that is not the case, and often molecular biosynthesis proceeds with a non-random distribution of the heavy isotopes, ending with the preferential formation of certain isotopomers versus others.

Relevant examples of early descriptions of this phenomenon are the synthesis of aliphatic amino acids (Bengsch and Grivet 1981), and glucose (Rossmann et al. 1991). The non-random distribution of ^{13}C during glucose synthesis results in ^{13}C enrichments in positions 1–3 as compared to 4–6 of the glucose molecule (Gilbert et al. 2009). This process is also at the origin of the ^{13}C natural abundance differences between specific C compounds. Lipids and lignin are depleted in ^{13}C compared to celluloses, and both are relatively depleted in ^{13}C compared to starch (Gleixner et al. 1993). Additionally, the distribution of ^{13}C across the six C positions of glucose produced by C3 and C4 plants show a consistent variation. C3 plants produce glucose enriched in ^{13}C (by ~2.6‰) in the C-3 position compared to C4 plants. C4 plants produce glucose enriched in ^{13}C in the C-1 position (by ~2.2‰) compared to that of C3 plants. Hobbie and Werner (2004) describe them as dependent on differences in photorespiration, sugar metabolism, and fluxes during the reductive pentose phosphate pathway between plants using C3 versus C4 photosynthesis.

The production of N_2O provides another example of processes that result in different isotopomers. N_2O can be produced by nitrification and denitrification processes driven by different functional microbial groups. Each process results in a distinct site-specific ^{15}N enrichment of the N_2O molecule (e.g., α position versus β position, $N^{\alpha}N^{\beta}O$; Box 2.1). This site preference ($SP = {}^{15}N^{\alpha} - {}^{15}N^{\beta}$) can thus be used to distinguish between different sources of N_2O (Baggs 2008).

References

Amundson, R., L. Stern, T. Baisden and Y. Wang (1998). "The isotopic composition of soil and soil-respired CO_2." *Geoderma* **82**(1–3): 83–114.

Andersen, K. K., N. Azuma, J. M. Barnola, M. Bigler, P. Biscaye, N. Caillon, J. Chappellaz, H. B. Clausen, D. Dahl-Jensen, H. Fischer, J. Flückiger, D. Fritzsche, Y. Fujii, K. Goto-Azuma, K. Grønvold, N. S. Gundestrup, M. Hansson, C. Huber, C. S. Hvidberg, S. J. Johnsen, U. Jonsell, J. Jouzel, S. Kipfstuhl, A. Landais, M. Leuenberger, R. Lorrain, V. Masson-Delmotte, H. Miller, H. Motoyama, H. Narita, T. Popp, S. O. Rasmussen, D. Raynaud, R. Rothlisberger, U. Ruth, D. Samyn, J. Schwander, H. Shoji, M. L. Siggard-Andersen, J. P. Steffensen, T. Stocker, A. E. Sveinbjörnsdóttir, A. Svensson, M. Takata, J. L. Tison, T. Thorsteinsson, O. Watanabe, F. Wilhelms, J. W. C. White and North Greenland Ice Core Project members (2004). "High-resolution record of Northern Hemisphere climate extending into the last interglacial period." *Nature* **431**(7005): 147–151.

Baggs, E. M. (2008). "A review of stable isotope techniques for N_2O source partitioning in soils: recent progress, remaining challenges and future considerations." *Rapid Communications in Mass Spectrometry* **22**(11): 1664–1672.

Bender, M. M. (1968). "Mass spectrometric studies of carbon 13 variations in corn and other grasses." *Radiocarbon* **10**(2): 468–472.

Bengsch, E. and J.-P. Grivet (1981). "Non-statistical label distribution in biosynthetic ^{13}C enriched amino acids." *Zeitschrift für Naturforschung B* **36**(10): 1289–1296.

Bigeleisen, J. (1965). "Chemistry of isotopes." *Science* **147**(3657): 463.

Brunner, B. and S. M. Bernasconi (2005). "A revised isotope fractionation model for dissimilatory sulfate reduction in sulfate reducing bacteria." *Geochimica et Cosmochimica Acta* **69**(20): 4759–4771.

Cerling, T. E., J. M. Harris, B. J. MacFadden, M. G. Leakey, J. Quade, V. Eisenmann and J. R. Ehleringer (1997). "Global vegetation change through the Miocene/Pliocene boundary." *Nature* **389**(6647): 153–158.

Coplen, T. B., J. A. Hopple, J. K. Böhlke, H. S. Peiser, S. E. Rieder, H. R. Krouse, K. J. R. Rosman, T. Ding, R. D. Vocke, Jr., K. M. Revesz, A. Lamberty, P. Taylor and P. De Bievre (2002). *Compilation of minimum and maximum isotope ratios of selected elements in naturally occurring terrestrial materials and reagents.* Water-Resources Investigations Report 01-4222. Reston, VA, U.S. Department of the Interior, U.S. Geological Survey.

Craig, H. (1954). "Carbon 13 in plants and the relationships between carbon 13 and carbon 14 variations in nature." *Journal of Geology* **62**(2): 115–149.

Craine, J. M., A. J. Elmore, L. Wang, L. Augusto, W. T. Baisden, E. N. J. Brookshire, M. D. Cramer, N. J. Hasselquist, E. A. Hobbie, A. Kahmen, K. Koba, J. M. Kranabetter, M. C. Mack, E. Marin-Spiotta, J. R. Mayor, K. K. McLauchlan, A. Michelsen, G. B. Nardoto, R. S. Oliveira, S. S. Perakis, P. L. Peri, C. A. Quesada, A. Richter, L. A. Schipper, B. A. Stevenson, B. L. Turner, R. A. G. Viani, W. Wanek and B. Zeller (2015). "Convergence of soil nitrogen isotopes across global climate gradients." *Scientific Reports* **5**: 8280.

Dawson, T. E., S. Mambelli, A. H. Plamboeck, P. H. Templer and K. P. Tu (2002). "Stable isotopes in plant ecology. " *Annual Review of Ecology and Systematics* **33**: 507–559.

Ehleringer, J. R. and T. E. Cerling (2002). "C3 and C4 photosynthesis." *Encyclopedia of Global Environmental Change* **2**(4): 186–190.

Farquhar, G. D. (1983). "On the nature of carbon isotope discrimination in C_4 species." *Functional Plant Biology* **10**(2): 205–226.

Farquhar, G. D., J. R. Ehleringer and K. T. Hubick (1989). "Carbon isotope discrimination and photosynthesis." *Annual Review of Plant Physiology and Plant Molecular Biology* **40**(1): 503–537.

Farquhar, G. D., M. H. O'Leary and J. A. Berry (1982). "On the relationship between carbon isotope discrimination and the intercellular carbon dioxide concentration in leaves." *Functional Plant Biology* **9**(2): 121–137.

Fry, B. (2006). *Stable isotope ecology.* New York, Springer.

Gat, J. R., A. Shemesh, E. Tziperman, A. Hecht, D. Georgopoulos and O. Basturk (1996). "The stable isotope composition of waters of the eastern Mediterranean Sea." *Journal of Geophysical Research: Oceans* **101**(C3): 6441–6451.

Gilbert, A., V. Silvestre, R. J. Robins and G. S. Remaud (2009). "Accurate quantitative isotopic ^{13}C NMR spectroscopy for the determination of the intramolecular distribution of ^{13}C in glucose at natural abundance." *Analytical Chemistry* **81**(21): 8978–8985.

Gleixner, G., H. J. Danier, R. A. Werner and H. L. Schmidt (1993). "Correlations between the ^{13}C content of primary and secondary plant products in different cell compartments and that in decomposing basidiomycetes." *Plant Physiology* **102**(4): 1287–1290.

Hobbie, E. A. and R. A. Werner (2004). "Intramolecular, compound-specific, and bulk carbon isotope patterns in C3 and C4 plants: a review and synthesis." *New Phytologist* **161**(2): 371–385.

Högberg, P. (1997). "Tansley review No 95. ^{15}N natural abundance in soil-plant systems." *New Phytologist* **137**(2): 179–203.

Horst, A., G. Lacrampe-Couloume and B. Sherwood Lollar (2016). "Vapor pressure isotope effects in halogenated organic compounds and alcohols dissolved in water." *Analytical Chemistry* 88(24): 12066–12071.

Mook, W. G., M. Koopmans, A. F. Carter and C. D. Keeling (1983). "Seasonal, latitudinal, and secular variations in the abundance and isotopic ratios of atmospheric carbon dioxide: 1. Results from land stations." *Journal of Geophysical Research: Oceans* **88**(C15): 10915–10933.

Mook, W. G. and K. Rozanski (2000). *Environmental isotopes in the hydrological cycle.: principles and applications*. Volumes I, IV and V. Paris, United Nations Educational, Scientific and Cultural Organization and International Atomic Energy Agency.

Morasch, B., H. H. Richnow, B. Schink and R. U. Meckenstock (2001). "Stable hydrogen and carbon isotope fractionation during microbial toluene degradation: mechanistic and environmental aspects." *Applied and Environmental Microbiology* **67**(10): 4842–4849.

Nier, A. O. and E. A. Gulbransen (1939). "Variations in the relative abundance of the carbon isotopes." *Journal of the American Chemical Society* **61**(3): 697–698.

O'Leary, M. H. (1993). Biochemical basis of carbon isotope fractionation. In: *Stable Isotopes and Plant Carbon-water Relations*. J. R. Ehleringer, A. E. Hall and G. D. Farquhar, eds. San Diego, CA, Academic Press: 19–28.

O'Leary, M. H., S. Madhavan and P. Paneth (1992). "Physical and chemical basis of carbon isotope fractionation in plants." *Plant, Cell & Environment* **15**(9): 1099–1104.

Pavlov, A. and J. Kasting (2002). "Mass-independent fractionation of sulfur isotopes in Archean sediments: strong evidence for an anoxic Archean atmosphere." *Astrobiology* **2**(1): 27–41.

Rayleigh, L. (1902). "LIX. On the distillation of binary mixtures." *The London, Edinburgh, and Dublin Philosophical Magazine and Journal of Science* **4**(23): 521–537.

Rees, C. E. (1973). "A steady-state model for sulphur isotope fractionation in bacterial reduction processes." *Geochimica et Cosmochimica Acta* **37**(5): 1141–1162.

Rossmann, A., M. Butzenlechner and H. L. Schmidt (1991). "Evidence for a nonstatistical carbon isotope distribution in natural glucose." *Plant Physiology* **96**(2): 609–614.

Schauble, E. A. (2004). "Applying stable isotope fractionation theory to new systems." *Reviews in Mineralogy and Geochemistry* **55**(1): 65–111.

Schlesinger, W. H. and E. S. Bernhardt (2013). *Biogeochemistry: an analysis of global change*. Amsterdam, Elsevier/Academic Press.

Scott, K., X. Lu, C. Cavanaugh and J. Liu (2004). "Optimal methods for estimating kinetic isotope effects from different forms of the Rayleigh distillation equation." *Geochimica et Cosmochimica Acta* **68**(3): 433–442.

Thiemens, M. and J. Heidenreich (1983). "The mass-independent fractionation of oxygen—a novel isotope effect and its possible cosmochemical implications." *Science* **219**(4588): 1073–1075.

Vogel, J. C., P. M. Grootes and W. G. Mook (1970). "Isotopic fractionation between gaseous and dissolved carbon dioxide." *Zeitschrift für Physik A: Hadrons and Nuclei* **230**(3): 225–238.

Vértes, A., S. Nagy, Z. Klencsár, R. G. Lovas, F. Rösch (2011). *Handbook of nuclear chemistry*. Boston, MA, Springer Science+Business Media B.V.

Wang, X. and H. A. J. Meijer (2018). "Ice–liquid isotope fractionation factors for 18O and 2H deduced from the isotopic correction constants for the triple point of water." *Isotopes in Environmental and Health Studies* **54**(3): 304–311.

Wickman, F. E. (1952). "Variations in the relative abundance of the carbon isotopes in plants." *Nature* **169**(4312): 1051–1051.

Activity: Calculating Measures of Isotopic Fractionation

Objectives

- Practice calculating fractionation factors and isotopic discrimination using ecological examples.
- Examine biogeochemical processes that lead to isotopic fractionation.

Tools

This activity can be completed entirely by hand with only a calculator. You may use a spreadsheet or data processing software if preferred. Refer to the equations for calculating fractionation factors and isotopic discrimination in Chapter 3 to help you complete the activities.

Activity

Consider the examples for C, N, and water cycles depicted below. Apply your understanding of the equations in Chapter 3 to calculate fractionation factors and isotopic discrimination. The values given are realistic but were made up for the purpose of this activity.

Exercise 1

Calculate fractionation factor and isotopic discrimination in the C cycle

Using the $\delta^{13}C$ data for the pools given in Figure A3.1, calculate the fractionation factor (α) and heavy isotope discrimination (Δ) associated with each process. Complete each calculation using both δ and R notation.

Exercise 2

Calculate fractionation factor and isotopic discrimination in meteoric water

Using the $\delta^{18}O$ and $\delta^{2}H$ data for meteoric water across the continental United States shown in Figure A3.2, calculate the fractionation factor (α) and heavy isotope discrimination (Δ) associated with continental transport of atmospheric water. Complete the calculations for each line with reference to the coastal δ value (least depleted), using both the δ and R notation.

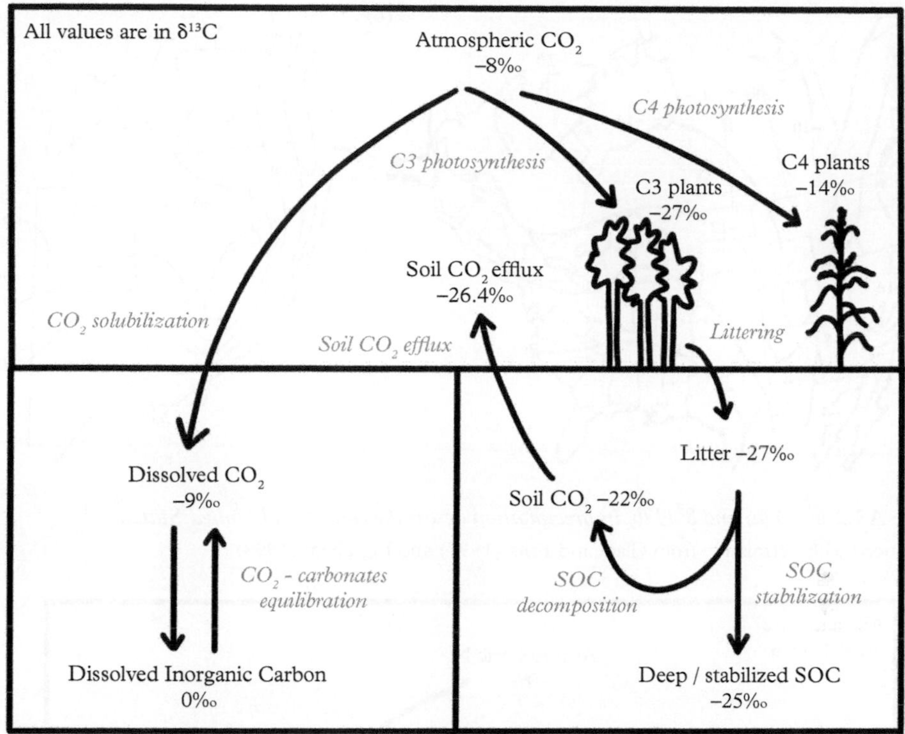

Figure A3.1 *Schematic of global C cycling pools (black font) with associated $\delta^{13}C$ values and processes (gray font, italicized). All values shown are in ‰ $\delta^{13}C$, they are realistic but created for the purpose of this exercise.*

Exercise 3

Calculate $\delta^{15}N$ based on fractionation factors

The schematic of the N cycle in Figure A3.3 is missing several $\delta^{15}N$ values. Using the given $\delta^{15}N$ and α values, calculate the missing $\delta^{15}N$ values.

Review questions

1. Compare the Δ values you calculated using Eq. 3.4 and Eq. 3.6 in Chapter 3. By how much does the isotopic fractionation differ between the two equations? Describe a scenario in which you could appropriately use each equation.
2. Describe the spatial pattern of ^{18}O isotopes in precipitation across the continental United States. Why does this pattern exist?
3. Which N cycling processes lead to the greatest isotopic fractionation? Propose a hypothesis for why this might be.

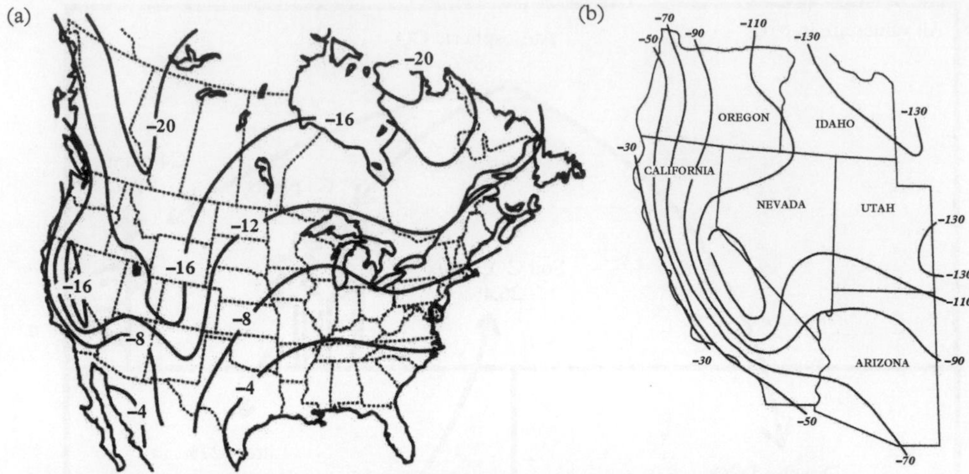

Figure A3.2 *$\delta^{18}O$ (a) and δ^2H (b) in precipitation across the continental United States.*
Reproduced with permission from Clark and Fritz (1997) and Ingraham (1998).

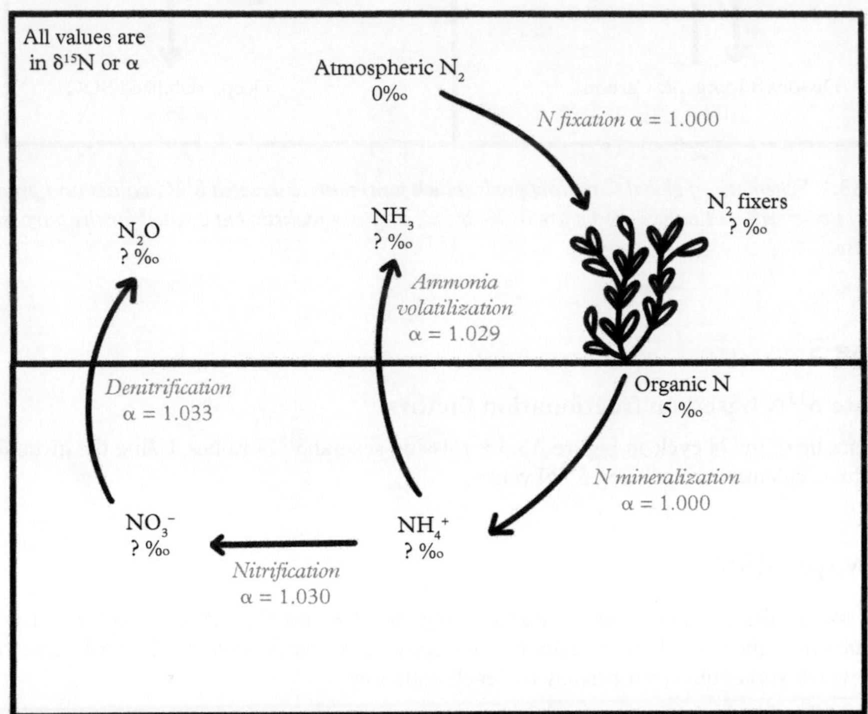

Figure A3.3 *Schematic of terrestrial N cycling pools (black font) with some $\delta^{15}N$ values (‰) given and processes (gray font, italicized) with α values given. The data are realistic but created for the purpose of this exercise.*

References

Clark, I. D. and P. Fritz (1997). *Environmental isotopes in hydrogeology*. Boca Raton, FL, CRC Press.

Ingraham, N. L. (1998). Isotopic variations in precipitation. In: *Isotope tracers in catchment hydrology*. C. Kendall and J. J. McDonnell, eds. Amsterdam, Elsevier: 87–118.

4

Isotope Mixing

4.1 Principles of mass balance

As an element flows from one pool to another in its biogeochemical cycle, multiple sources of the element mix into these pools, but the element preserves its mass. Imagine, for example, the pool of water vapor in the atmosphere. It is continuously replenished by the evaporation of water from multiple sources, such as the oceans, freshwaters, and the transpiration of plants. The resulting atmospheric water vapor pool will have a mass proportional to the relative contribution (f) of the source pools multiplied by their respective water masses. In mathematical terms, starting with the simplest case when two pools (A and B) mix to form one sample (S), the resulting sample will have a mass (m_S) proportional to the sum of the masses of the two contributing pools:

$$m_S = m_A + m_B \tag{4.1}$$

and the fractional contribution of one pool (f_A) will correspond to the ratio of the masses of that pool over the total:

$$f_A = m_A/m_S \tag{4.2}$$

The contribution of the second pool (f_B) will be:

$$f_B = 1 - f_A \tag{4.3}$$

This conservation of mass in biogeochemical cycling provides the opportunity to use isotopes to quantify the relative contribution of one or more pools to a mixture, applying what is well known as the *mass balance equation* or *isotopic mixing model*. As we learned in Chapter 3, fractionation generates variations in the isotopic composition of natural pools of an element (see Section 3.7). If two such pools with distinct enough isotopic composition form a mixture, an isotopic mixing model can be applied to partition each pool's contribution to that mixture. As we will learn in Chapter 5, when natural abundance differences are not large enough, scientists often manipulate the isotopic compositions of pools, in order to create the conditions to apply the isotopic mixing model. Either way, source partitioning using isotopic mixing models is probably the most common application of isotopes in ecology! Let's understand how it works.

A Primer on Stable Isotopes in Ecology. M. Francesca Cotrufo and Yamina Pressler, Oxford University Press.
 DOI: 10.1093/oso/9780198854494.003.0004

The isotopic composition of a pool affects its mass. Therefore, the mass balance Eq. 4.1 can be rewritten as:

$$\delta_S \cdot m_S = \delta_A \cdot m_A + \delta_B \cdot m_B \tag{4.4}$$

where δ_S, δ_A, and δ_B are the isotopic composition in δ notation of the mixture pool S, and the contributing pools A and B, respectively. The isotopic composition of the mixture pool (S) can then be calculated as:

$$\delta_S = (\delta_A \cdot m_A + \delta_B \cdot m_B) / (m_A + m_B) \tag{4.5}$$

which means that the isotopic composition of a mixture pool (δ_S) is proportional to the isotopic composition of its contributing pools (δ_A and δ_B), weighted by their contribution to the mixture pool. It is worth noticing that in this chapter we are using δ notations for simplicity, but the same applies for fractional or atom % notations, which are actually the exact notations to use in isotopic mixing models when the heavy isotope abundance exceeds 10 atom % (see Section 2.5).

Combining and rewriting the above equations, we obtain the most commonly used expression of an isotopic mixing model:

$$f_A = (\delta_S - \delta_B) / (\delta_A - \delta_B) \tag{4.6}$$

Why is this expression so commonly used? It allows researchers to determine the relative contributions (f) of two sources to the mixture by measuring the isotopic composition of the mixture (δ_S) and the two contributing pools (δ_A and δ_B), which are commonly defined as "end members" in isotopic mixing models. The *f value* can only vary between 0 and 1, and therefore the δ value of the mixture pool must be within the range of the δ values of the two end members (Figure 4.1). If this principle doesn't apply, at least one of the two end-member pools did not contribute to the mixture. Checking if end member δ values are distinct enough (see Section 4.7) and that the δ value of the mixture falls within them is the first step to deciding whether or not a mixing model can be applied in a given system.

Isotope studies often discuss a pool's contribution to a mixture in terms of f values. If the mass of the mixture pool is also measured, the actual mass contribution of a pool to the mixture can be calculated as:

$$m_A = f_A \cdot m_S \tag{4.7}$$

The beauty of mass balance equations is that they allow us to use isotopes for quantitative studies of biogeochemical cycling!

4.2 Applying the two end-member mixing model

To avoid getting lost in the equations above, let's consider a few examples to clarify how the isotopic mixing model works and why it is so useful.

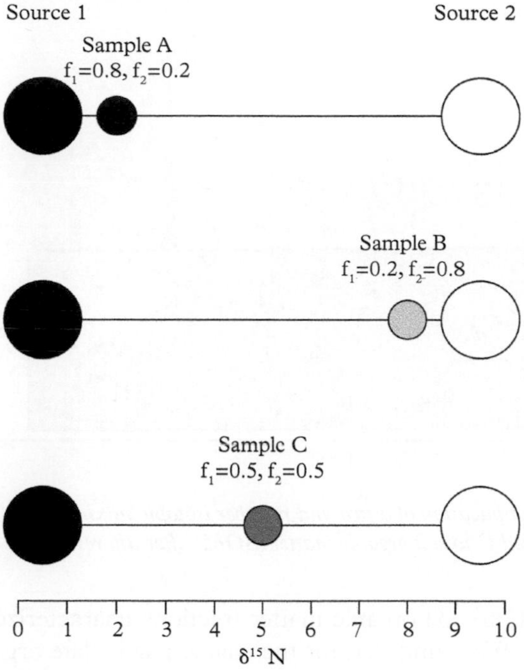

Figure 4.1 *Schematic of a two-pool mixing model. Source 1 contributes f_1 and source 2 contributes f_2 to each sample. Sources contribute unequally to samples A and B, and contribute equally to sample C.* Modified from Fry (2006).

We are soil ecologists, and one of the pressing needs in our discipline is to understand how to store C in the soil, to remove it from the atmosphere where it contributes to climate warming. Thus, tracing C from its sources (e.g., plant inputs) into the soil is a core activity of our research, and an area in which we commonly apply isotopes. We learned in Chapter 3 that plants with a C3 photosynthetic pathway form organic matter with a lower ^{13}C abundance than plants with a C4 photosynthetic pathway (see Figure 3.7). Thus, if trees (C3) are planted on fields historically cultivated under maize (a C4 grass), the relative contribution of tree-derived C (new C) and maize-derived C (old C) to the soil organic matter C (SOM C) can be calculated knowing the $\delta^{13}C$ of the two contributing pools (i.e., tree-derived (δ_{new}) and maize-derived (δ_{old}) organic matter) and of the mixture (i.e., soil organic matter (δ_{SOM})), applying the isotopic mixing model (Figure 4.2):

$$f_{new} = (\delta_{SOM} - \delta_{old}) / (\delta_{new} - \delta_{old}) \quad (4.8)$$

That's what we did a few years ago, to understand the effects of 20 years of afforestation on C storage in soil previously cultivated with maize (Del Galdo et al. 2003). In that study, we sourced the contribution of new (tree-derived, $\delta_{new} = -27.17‰$) and

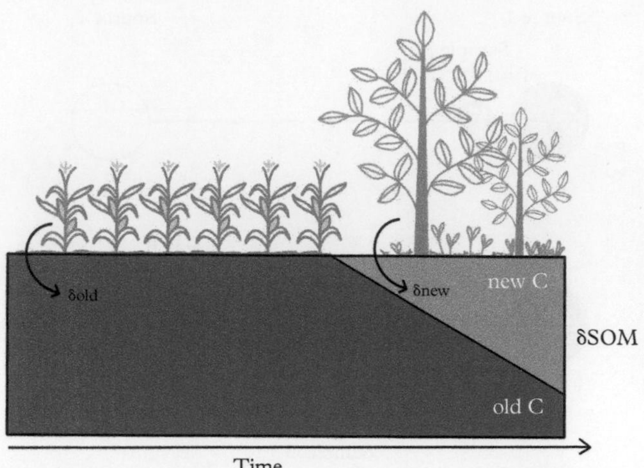

Figure 4.2 *Example of application of a two-end member isotopic mixing model to determine the contribution of tree-derived C to soil organic matter (SOM) after afforestation.*

old (maize-derived) C to soil organic matter fractions characterized by different time of persistence in soil. We found that for the coarse particulate organic matter δ_{SOM} = −25.96‰ and δ_{old} = −19.75‰, resulting in:

$$f_{new} = (-25.96 + 19.75) / (-27.17 + 19.75) = 0.84 \tag{4.9}$$

meaning that 84% of the coarse particulate organic matter C was derived from tree inputs after 20 years of afforestation, testifying to the faster turnover of this pool. By contrast, for the mineral associated organic matter fraction δ_{SOM} = −23.05‰ and δ_{old} = −21.56‰, resulting in:

$$f_{new} = (-23.05 + 21.56) / (-27.17 + 21.56) = 0.26 \tag{4.10}$$

meaning that only 26% of the mineral-associated organic matter C was derived from the tree inputs after 20 years of afforestation, demonstrating the slower turnover of this pool.

The two end-member isotopic mixing model has been applied to a large variety of ecological and biogeochemical studies, from the early reconstruction of grass–tree savanna ecotones (Martin et al. 1990), to sourcing root water (Dawson and Pate 1996), and many others the reader can find by searching the literature.

4.3 Accounting for fractionation in mixing models

So far, we have discussed isotope mixing as if there were no fractionation associated with the movement of isotopes from the source end members to the mixture. However, as we learned in Chapter 3, many processes discriminate against heavy or light isotopes. How do we account for isotopic fractionation in mixing models?

The simplest way to account for fractionation is to identify end members of the same pool as the mixture, rather than of the actual sources, but that represent sources' contributions to the mixture. What does that mean in practical terms?

Let's go back to our previous example where we wanted to determine the amount of soil organic C derived from tree C versus that derived from maize C. While soil organic C mostly maintains the isotopic signal of the plant material from which it is derived, it slightly enriches in ^{13}C (Balesdent et al. 1993). This difference is due to microbial transformations during decomposition, some possible carboxylation reactions within the soil, and other processes (Ehleringer et al. 2000). For this reason, in cases when fractionation is suspected, *the most correct mixing model uses end members that are the same pools as the mixture* (i.e., soil organic C), *but represent contributions by the individual sources* (e.g., soil organic matter formed exclusively from maize (δ_{old}), rather than maize tissues directly). This is not always possible, but when it is possible, it is, in our opinion, the best approach to use since any type of isotope fractionation can be ignored!

An alternative approach requires having an accurate estimate of the fractionation factor associated with the transformation of the source pools to the mixture pool. Here, we must correct for the fractionation in the mixing model expression. This is, for example, the case when partitioning food sources of animal diets, in which case appropriate isotopic fractionation factors associated with animal assimilation are required for mixing model corrections (Caut et al. 2009). In this case, in δ values, the mixing model becomes:

$$f_A = [(\delta_S - \Delta) - \delta_B] / (\delta_A - \delta_B) \tag{4.11}$$

where f_A is the fractional contribution of pool A and δ_A and δ_B are the isotopic composition of the two sources (i.e., the end members of pools A and B). δ_S is the isotopic composition of the mixture (sample) from which the isotopic enrichment (Δ) associated with the mixing process is subtracted, if it is deemed to be the same for both sources.

If two sources are known to fractionate differently, then the mixing model becomes:

$$f_A = [\delta_S - (\delta_B + \Delta_B)] / [(\delta_A + \Delta_A) - (\delta_B + \Delta_B)] \tag{4.12}$$

with Δ_A and Δ_B representing the heavy isotopic enrichment associated with the transformation of the pool A to S and the pool B to S, respectively. This, of course, is most accurately done when the heavy isotopic enrichment (Δ) values are experimentally determined. Alternatively, when Δ values from the literature are used, a sensitivity analysis examining the effect of variation in Δ is required (Wolf et al. 2009), making this approach extremely difficult to use.

4.4 Concentration-dependent isotope mixing

In biogeochemistry, we most often follow an element as it cycles from one pool into another. Therefore, the most common isotopic mass balance expression used is as presented in Eq. 4.4. However, there are cases in which we may be interested in tracing how much an entire pool (not just its element) contributes to the formation of a new pool. In this case, the elemental concentration in the source pools and the mixture need to be taken into account in the isotopic mass balance, which becomes:

$$\delta_S = (\delta_A \cdot m_A \cdot C_A + \delta_B \cdot m_B \cdot C_B) / (m_A \cdot C_A + m_B \cdot C_B) \quad (4.13)$$

where C_A and C_B are the concentrations of the element of interest in the two end-member pools A and B, respectively. Accounting for concentration is critical when the two end members have very different elemental concentrations, and thus their mass contribution to the mixture would be erroneously calculated if concentrations were not accounted for. As depicted in Figure 4.3, when the two end-member pools have the same concentration (1:1), their fractional contribution to the mixture varies linearly with the δ of the mixture pool. However, the greater the elemental concentration of one end-member pool is than the other, the more curved the mixing relationship becomes, flexing toward a higher fractional contribution of the pool with the highest concentration.

In practical terms, when does an isotope ecologist need to apply a concentration-dependent mixing model? A biogeochemist almost never applies it because we typically trace the fate of individual elements (e.g., C, N, O, and H) as they flow from plants to microorganisms, soil, the atmosphere, etc. However, nutritional isotope ecologists, for example, who want to know how much an animal diet depends on another animal versus

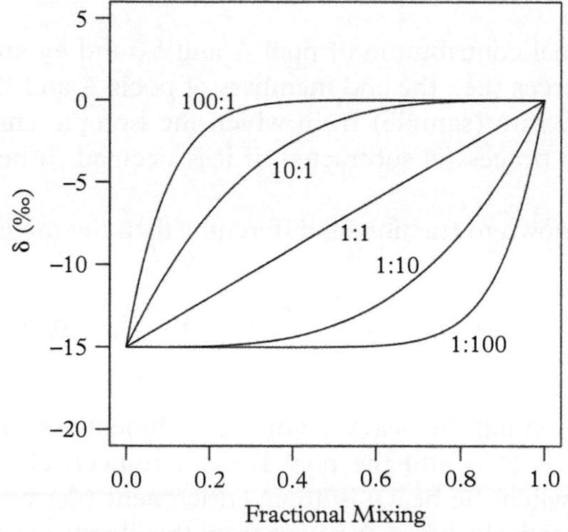

Figure 4.3 *Fractional mixing when end-member concentrations differ. Examples for end-member concentrations from 1:1, 1:10, and 1:100 and δ values of 0 and −15‰ are displayed.*
Modified from Clark and Fritz (1997).

a plant, using the natural ^{15}N abundance variation between these two sources, will need to take into account that the elemental N concentration is much higher in animal than plant tissues, or else will come to an erroneous conclusion. Concentration-dependent mixing models using N and O stable isotopes have also been applied to quantify nitrate (NO_3^-) inputs into river water (Deutsch et al. 2006). Phillips and Koch (2002) provide a great illustration and a robust discussion of concentration dependence in mixing models.

4.5 From mass balance to the Keeling plot

In the previous sections we discussed the use of mass balance to determine the relative contribution of one of two sources to a mixture, when the isotope composition of the two end members (i.e., sources) and the mixture pool are known. What about when one of the two end members' isotopic composition is the unknown? Can we use the mass balance approach to calculate it?

The answer is yes, and the approach is commonly known as the "Keeling plot" approach, from Charles D. Keeling who developed it when studying the daily cycle of ^{13}C and ^{18}O abundance in atmospheric CO_2 due to the varying contribution of photosynthesis and respiration across the day, in rural areas along the Pacific coast of North America (Keeling 1958).

The Keeling plot approach is used to determine the isotopic composition of a source when its contribution to the mixture changes over time and the mass and isotopic composition of the mixture can be quantified. Mathematically, the Keeling plot derives from the mass balance equation, where:

$$\delta_S \cdot m_S = \delta_A \cdot m_A + \delta_B \cdot m_B \tag{4.14}$$

is the mass balance equation, and rearranging:

$$\delta_S = \delta_B + (1/m_S) \cdot m_A \cdot (\delta_A - \delta_B) \tag{4.15}$$

we obtain a linear equation of the kind:

$$Y = a + 1/x \cdot b \tag{4.16}$$

where our unknown (δ_B, the isotopic composition of one of two end members) is the intercept (a), in the relationship between the isotopic composition of the mixture (δ_S) and the inverse of its mass ($1/m_S$), as illustrated in Figure 4.4.

In terrestrial biogeochemistry, Keeling plots are commonly used to determine the isotopic composition of carbon and water fluxes into the atmosphere at multiple scales (Pataki et al. 2003; Williams et al. 2004). For example, the $\delta^{13}C$ of the ecosystem respiration flux can be determined by monitoring the changes in atmospheric CO_2 concentration and its ^{13}C abundance over time in the dark (to avoid interference with photosynthetic uptake). As the ecosystem respiration adds CO_2 to the atmosphere, the CO_2 concentration will increase while the $\delta^{13}C$ will decrease because plants and soil organic matter have a lower ^{13}C natural abundance than the atmosphere (Bowling et al. 2008). Thus, plotting the $\delta^{13}C$-CO_2 against the inverse of the concentration we obtain a linear relationship with the intercept representing our unknown (i.e., the $\delta^{13}C$ of the ecosystem respiration), and the $\delta^{13}C$-CO_2 of the free atmosphere representing the other end member (Figure 4.5).

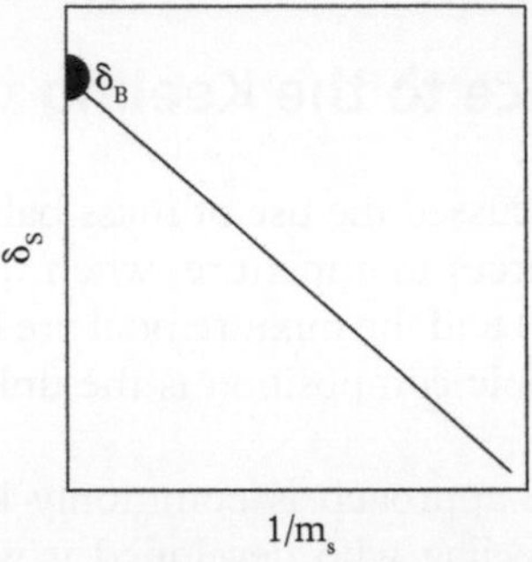

Figure 4.4 *Schematic demonstrating an example of the relationship between the isotopic composition of the unknown (δ_B), the isotopic composition of the mixture (δ_S), and the inverse of its mass ($1/m_S$). The slope of the line can vary depending on the system.*

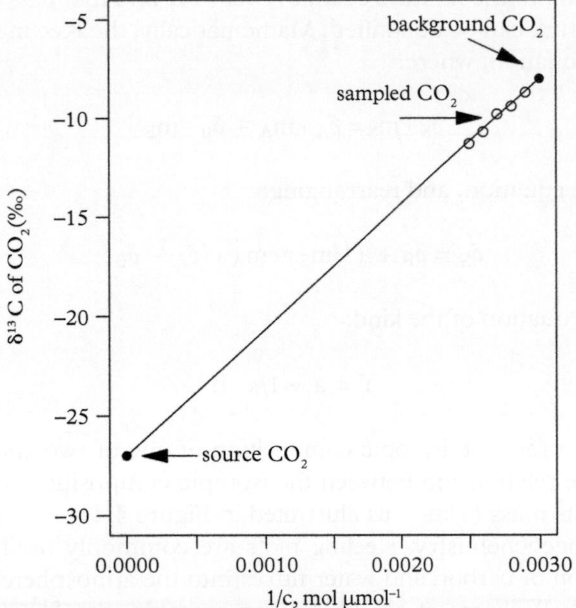

Figure 4.5 *Illustration of the Keeling plot method showing the $\delta^{13}C$ of the two end members (source CO_2 and background atmospheric CO_2; closed circles) and the mixture (sampled CO_2; open circles).* Modified from Pataki et al. (2003). Figure has been recreated as an approximation to show the concept but does not reflect the original data.

The linear regression used in standard Keeling plots assumes that the independent variable ($1/m_S$) either has no error or that errors are under the control of the experimenter, and they are independent from the error associated with the dependent variable (δ_S). Since this assumption is not met, alternative formulations that take the errors on $1/m_S$ and δ_S into account can be used. Pataki et al. (2003, and references therein) discuss this issue extensively and illustrate how the geometric mean regression model can be used in place of the linear model for a more correct expression of the Keeling plot.

4.6 Three-source mixing models

Mixing models are straightforward to use when there are only two sources which contribute to the mixture. What if there are more sources?

If there are three sources (A, B, and C), then there are two unknowns (e.g., f_A, f_B), from which the third can be calculated [$f_C = 1 - (f_A + f_B)$]. Clearly, we need a system with three equations to solve for two unknowns. That is, we need two isotopes! If the three sources are sufficiently different for two isotopes (X and Y), then their relative contribution to the mixture (S) can be calculated from the following system of equations:

$$\begin{aligned} \delta X_S &= f_A\delta X_A + f_B\delta X_B + f_C\delta X_C \\ \delta Y_S &= f_A\delta Y_A + f_B\delta Y_B + f_C\delta Y_C \\ f_A &+ f_B + f_C = 1 \end{aligned} \tag{4.17}$$

When three sources contribute to a mixture, the mixture will fall inside the triangle defined by the δx and δy values of the three sources (Figure 4.6). If that does not occur, one or more of the sources did not contribute to the mixture, or some other issues had not been accounted for, such as the concentration dependency of the relative contribution of the sources, or different fractionation factors for the different sources (Phillips and Koch 2002).

A three-pool isotopic mixing model was used, for example, by Oelmann and collaborators (2007). They wanted to identify the contribution to the NO_3^- mixture pool (M) in a soil solution of rainfall (R), and of mineralization from soil organic matter derived from leguminous (Leg) and from non-leguminous (SOM) inputs. Since NO_3^- produced from each of these three sources (i.e., R, Leg, and SOM) had a distinct $\delta^{15}N$ and $\delta^{18}O$ from the others, they could apply a three-pool isotopic mixing model to address their question. They measured the $\delta^{15}N$ and $\delta^{18}O$ of each of the sources and of the soil NO_3^- pool (M), and created a system of mass balance equations as follow:

$$\begin{aligned} \delta^{15}N_M &= f_R\delta^{15}N_R + f_{Leg}\delta^{15}N_{Leg} + f_{SOM}\delta^{15}N_{SOM} \\ \delta^{18}O_M &= f_R\delta^{18}O_R + f_{Leg}\delta^{18}O_{Leg} + f_{SOM}\delta^{18}O_{SOM} \\ f_R &+ f_{Leg} + f_{SOM} = 1 \end{aligned} \tag{4.18}$$

Where subscripts denote the NO_3^- mixture (M), and the three sources (R, Leg, and SOM), and f represents their relative contribution to the mixture. The $\delta^{15}N$ and $\delta^{18}O$ values of NO_3^- in rainfall ($\delta^{15}N$: 3.3 ± 0.8‰; $\delta^{18}O$: 30.8 ± 4.7%), and from the

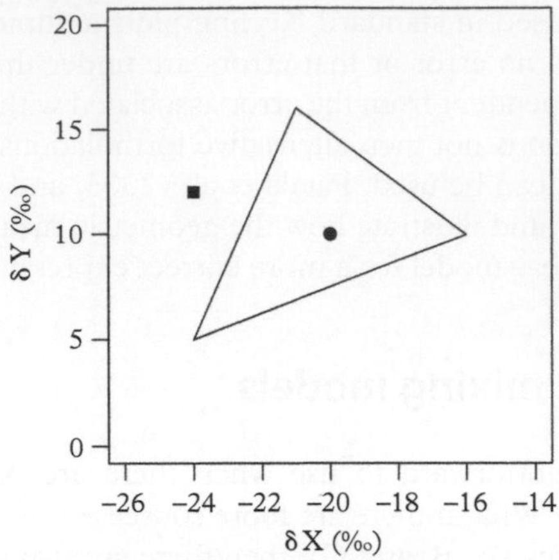

Figure 4.6 *Theoretical example of a three-pool mixing model showing a mixture that falls within the triangle defined by δX and δY of the three sources (filled circle), and a mixture that falls outside the triangle (filled square). One or more of the sources did not contribute to the square mixture because it falls outside of the source triangle.*

mineralization of Leg ($\delta^{15}N$: 9.3 ± 0.9%; $\delta^{18}O$: 6.7 ± 0.8%), and the mineralization of SOM ($\delta^{15}N$: 1.5 ± 0.6%; $\delta^{18}O$: 5.1 ± 0.9%) were markedly different. Applying the three-pool linear mixing model, they calculated 18–41% of the soil NO_3^- pool originated from rainfall, 38–57% from mineralization of non-leguminous SOM, and 18–40% from mineralization of leguminous SOM (Oelmann et al. 2007).

4.7 Uncertainty in source partitioning

Stable isotope mixing models are commonly used in a large variety of disciplines and applications. While it is tempting to apply them every time isotopic values of potential sources and their mixture are available, their blind application can often generate erroneous conclusions. Before using any mixing model, it is important to identify whether it can be used appropriately, and evaluate the uncertainty associated with the results it generates.

For linear mixing models, several factors may affect the uncertainty around the estimates of source partitioning. In their seminal paper, Phillips and Gregg (2001) conducted a sensitivity analysis of variation in source partition estimates in response to various factors, such as isotopic difference between sources, standard deviation of the isotopic values of end-members and mixture, sample size, analytical standard deviation, and evenness of the source proportion (Figure 4.7).

The isotopic difference between end members is the first most important determinant of the uncertainty associated with a linear mixing model. In a two end-member mixing

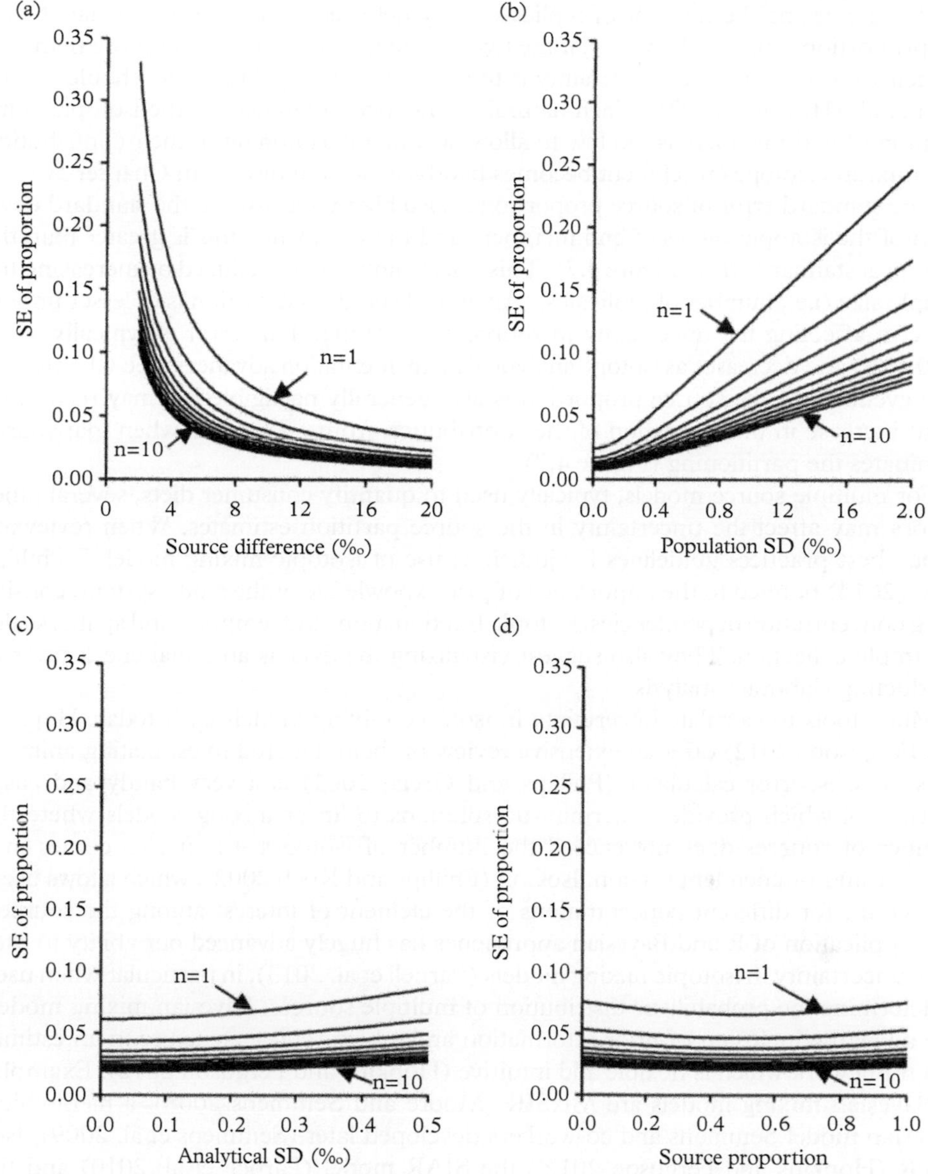

Figure 4.7 *Sensitivity analysis of variation in source partition estimates in response to isotopic difference between sources (a), standard deviation (SD) of the isotopic values of end members and mixture (b), analytical SD (c), and evenness of the source proportion (d). SE, standard error.* Reproduced with permission from Phillips and Gregg (2001).

model, increasing the number of replicates may help reduce the error associated with the proportion estimate. However, if the two end-members do not differ more than 5‰ in their isotopic signature, uncertainty is too high and the mixing model should not be used at all (Högberg 1997). When natural abundance differences in the isotopic composition of end members is too low to allow accurate partitioning of their contribution to a mixture, isotope enrichment becomes handy, as we will discuss in Chapter 5.

The standard error of source proportion is also highly sensitive to the standard deviation of the isotopic values of end members and mixture, when this is greater than the analytical standard error (Figure 4.7). This uncertainty can be reduced by increasing the sample size (i.e., number of replicates). Analytical standard deviation is of lesser importance in affecting the uncertainty in source partitioning. This error is typically small (0–0.5‰) and decreases as isotope analytical instrumentation advances (see Chapter 6). The evenness of the source proportion is also generally negligible but may result in a slight increase in the estimation of the contribution from one source when that source dominates the partitioning (Figure 4.7).

For multiple source models, typically used to quantify consumer diets, several other factors may affect the uncertainty in the source partition estimates. When reviewing some "best practices guidelines for judicious use of isotopic mixing models," Phillips et al. (2014) pointed to the importance of prior knowledge of the study system, considering concentration dependencies, isotopic fractionation, and temporal and spatial scales of sample collection. They also suggest visualizing the data as an initial check prior to conducting elaborate analysis.

Many tools to calculate uncertainty in isotopic mixing models exist today. Hopkins and Ferguson (2012) offer an extensive review of them, tailored to estimating animals' diets. The isoerror calculator (Phillips and Gregg 2001) is a very handy and easy-to-use tool which provides deterministic solutions to linear mixing models where the number of sources does not exceed the number of isotopes +1. It also comes in a concentration-dependent version, IsoCon (Phillips and Koch 2002), which allows users to account for different concentrations of the element of interest among the sources. The application of R and Bayesian approaches has hugely advanced our ability to estimate uncertainty in isotopic mixing models (Parnell et al. 2013), in particular when used to determine the probabilistic distribution of multiple sources. Bayesian mixing models also allow the inclusion of prior information and a hierarchical structure in an estimation framework which is flexible and intuitive (Hopkins and Ferguson 2012). Examples of Bayesian mixing models are MixSIR (Moore and Semmens 2008), a hierarchical Bayesian model Semmens and co-workers developed later (Semmens et al. 2009), IsotopeR (Hopkins and Ferguson 2012), the SIAR model (Parnell et al. 2010) and the later SIMMR (Parnell et al. 2013). Bayesian mixing models have been shown to outperform linear mixing models, as in the case of plant water source partitioning (Wang et al. 2019). As the field of modeling uncertainty will continue to evolve, we recommend staying informed with recent advancements and to use the tools that are most appropriate to the specific research questions, experimental design, and use of mixing models.

In many research settings, mixing models are used to obtain a source partition estimate, commonly referred to as the *f value*, which is subsequently used for further calculations. In these cases, rather than calculating the uncertainty associated with an average *f value*, and then having to apply the propagation of error to the following calculations, it can be more convenient to calculate an *f value* for each replicate, continue calculations by replicates, and calculate uncertainty on the final value of interest. One can do so by creating a row in the spreadsheet or data frame for each replicate and a column to enter a mixing model calculation for each replicate. Then, the solution (*f value*) obtained for each replicate can be used in following calculations by replicate (i.e., rows on your spreadsheet or data frame). When mixing models are independently applied at the replicate level, it is important to design the experiment accordingly. A design may follow a paired block scheme where each end-member replicate has its pair within a block, resulting in independent replicated mixing models (e.g., Soong et al. 2016). When this cannot be realized, average end-member values can be used (e.g., Stewart et al. 2013). In this case, replicate mixing models will vary for the mixture value, but will have the same average values for one or more end members. To date, the majority of the studies using linear mixing models for biogeochemical applications, use this approach and calculate and treat *f values* at the replicate level.

References

Balesdent, J., C. Girardin and A. Mariotti (1993). "Site-related δ(13)C of tree leaves and soil organic matter in a temperate forest." *Ecology* 74(6): 1713–1721.

Bowling, D. R., D. E. Pataki and J. T. Randerson (2008). "Carbon isotopes in terrestrial ecosystem pools and CO2 fluxes." *New Phytologist* **178**(1): 24–40.

Caut, S., E. Angulo and F. Courchamp (2009). "Variation in discrimination factors ($\Delta^{15}N$ and $\Delta^{13}C$): the effect of diet isotopic values and applications for diet reconstruction." *Journal of Applied Ecology* **46**(2): 443–453.

Clark, I. D. and P. Fritz (1997). *Environmental isotopes in hydrogeology*. Boca Raton, FL, CRC Press.

Dawson, T. E. and J. S. Pate (1996). "Seasonal water uptake and movement in root systems of Australian phraeatophytic plants of dimorphic root morphology: a stable isotope investigation." *Oecologia* **107**(1): 13–20.

Del Galdo, I., J. Six, A. Peressotti and M. F. Cotrufo (2003). "Assessing the impact of land-use change on soil C sequestration in agricultural soils by means of organic matter fractionation and stable C isotopes." *Global Change Biology* **9**(8): 1204–1213.

Deutsch, B., M. Mewes, I. Liskow and M. Voss (2006). "Quantification of diffuse nitrate inputs into a small river system using stable isotopes of oxygen and nitrogen in nitrate." *Organic Geochemistry* **37**(10): 1333–1342.

Ehleringer, J. R., N. Buchmann and L. B. Flanagan (2000). "Carbon isotope ratios in belowground carbon cycle processes." *Ecological Applications* **10**(2). 412–422.

Fry, B. (2006). *Stable isotope ecology*. New York, Springer.

Hopkins, J. B., III and J. M. Ferguson (2012). "Estimating the diets of animals using stable isotopes and a comprehensive Bayesian mixing model." *PLoS One* 7(1): e28478.

HÖGberg, P. (1997). "Tansley Review No. 95. ^{15}N natural abundance in soil-plant systems." *New Phytologist* **137**(2): 179–203.

Keeling, C. D. (1958). "The concentration and isotopic abundances of atmospheric carbon dioxide in rural areas." *Geochimica et Cosmochimica Acta* **13**(4): 322–334.

Martin, A., A. Mariotti, J. Balesdent, P. Lavelle and R. Vuattoux (1990). "Estimate of organic matter turnover rate in a savanna soil by ^{13}C natural abundance measurements." *Soil Biology and Biochemistry* **22**(4): 517–523.

Moore, J. W. and B. X. Semmens (2008). "Incorporating uncertainty and prior information into stable isotope mixing models." *Ecology Letters* **11**(5): 470–480.

Oelmann, Y., Y. Kreutziger, R. Bol and W. Wilcke (2007). "Nitrate leaching in soil: tracing the NO_3- sources with the help of stable N and O isotopes." *Soil Biology and Biochemistry* **39**(12): 3024–3033.

Parnell, A. C., R. Inger, S. Bearhop and A. L. Jackson (2010). Source partitioning using stable isotopes: coping with too much variation. *PLoS One* **5**(3):e9672.

Parnell, A., D. L. Phillips, S. Bearhop, B. Semmens, E. Ward, J. Moore, A. Jackson, J. Grey, D. Kelly, R. Inger (2013). "Bayesian stable isotope mixing models." *Environmetrics* **24**(6): 387–399.

Pataki, D. E., J. R. Ehleringer, L. B. Flanagan, D. Yakir, D. R. Bowling, C. J. Still, N. Buchmann, J. O. Kaplan and J. A. Berry (2003). "The application and interpretation of Keeling plots in terrestrial carbon cycle research." *Global Biogeochemical Cycles* **17**(1): 1022.

Phillips, D. L. and J. W. Gregg (2001). "Uncertainty in source partitioning using stable isotopes." *Oecologia* **127**(2): 171–179.

Phillips, D. L. and P. L. Koch (2002). "Incorporating concentration dependence in stable isotope mixing models." *Oecologia* **130**(1): 114–125.

Phillips, D. L., R. Inger, S. Bearhop, A. L. Jackson, J. W. Moore, A. C. Parnell, B. X. Semmens and E. J. Ward (2014). "Best practices for use of stable isotope mixing models in food-web studies." *Canadian Journal of Zoology* **92**(10): 823–835.

Semmens, B. X., E. J. Ward, J. W. Moore and C. T. Darimont (2009). "Quantifying inter- and intra-population niche variability using hierarchical bayesian stable isotope mixing models." *PLoS One* **4**(7): e6187.

Soong, J. L., M. L. Vandegehuchte, A. J. Horton, U. N. Nielsen, K. Denef, E. A. Shaw, C. M. de Tomasel, W. Parton, D. H. Wall and M. F. Cotrufo (2016). "Soil microarthropods support ecosystem productivity and soil C accrual: evidence from a litter decomposition study in the tallgrass prairie." *Soil Biology and Biochemistry* **92**: 230–238.

Stewart, C. E., J. Zheng, J. Botte and M. F. Cotrufo (2013). "Co-generated fast pyrolysis biochar mitigates green-house gas emissions and increases carbon sequestration in temperate soils." *GCB Bioenergy* **5**(2): 153–164.

Wang, J., N. Lu and B. Fu (2019). "Inter-comparison of stable isotope mixing models for determining plant water source partitioning." *Science of The Total Environment* **666**: 685–693.

Williams, D. G., W. Cable, K. Hultine, J. C. B. Hoedjes, E. A. Yepez, V. Simonneaux, S. Er-Raki, G. Boulet, H. A. R. de Bruin, A. Chehbouni, O. K. Hartogensis and F. Timouk (2004). "Evapotranspiration components determined by stable isotope, sap flow and eddy covariance techniques." *Agricultural and Forest Meteorology* **125**(3): 241–258.

Wolf, N., S. A. Carleton and C. Martínez Del Rio (2009). "Ten years of experimental animal isotopic ecology." *Functional Ecology* **23**(1): 17–26.

Activity: Applying Isotope Mixing Models to Ecology

Objectives

- Apply two- and three-source isotope mixing models to answer ecological questions.
- Apply the Keeling plot method to determine unknown isotopic compositions.

Tools

The solution to two-source mixing models can be calculated by hand with the use of a calculator, but we recommend using data processing software. For the three-source mixing model we suggest the use of available spreadsheets (e.g., isoerror calculators; Phillips & Gregg (2001)) or other data processing software and packages. Refer to Chapter 4 for selecting, solving, and interpreting isotope mixing models to help you complete the activities.

Exercise 1

Partition sources of soil respiration

Let's imagine you are studying soil C dynamics in three different ecosystems: (1) a chaparral, (2) a pasture, and (3) a wood plantation that was established on a former corn field. You are interested in understanding patterns of autotrophic (i.e., root respiration) and heterotrophic (i.e., respiration derived from soil organic matter only) respiration across the three ecosystems. You are also interested in quantifying the relative contribution to the total soil respiration flux, where possible.

To measure the total soil respiration (i.e., autotrophic + heterotrophic) flux and its isotopic composition, you deployed a soil collar at each site that allowed you to measure CO_2 concentrations and $\delta^{13}C\text{-}CO_2$ as the CO_2 accumulated in the chamber headspace. You repeated the CO_2 concentration and $^{13}C\text{-}CO_2$ measurements at eight points in time, and recorded the values (Table A4.1, field measurements)

To obtain the isotopic value of heterotrophic and autotrophic respiration (i.e., your two end members; see Sections 4.1 and 4.2), you also collected four replicate soil cores at each site, around the respiration collar. Back in the laboratory, you sieved the soil and separated the roots from the soil. Then, you rewetted each of the four replicates of soil without roots and root samples and incubated then in separate jars, by replicate. You measured the $\delta^{13}C\text{-}CO_2$ using gas chromatography–isotope ratio mass spectrometry (GC-IRMS) or a laser in each jar headspace, to determine the average and standard deviation ($n = 4$) of the $\delta^{13}C\text{-}CO_2$ of the heterotrophic (soil only) and autotrophic (root) respiration fluxes, and noted the results (Table A4.1, laboratory incubation).

Table A4.1 *Results of soil carbon field and laboratory measurements. Values are realistic but made up for the purpose of this exercise.*

Ecosystem		Field measurements			Laboratory incubation	
	Sampling times	CO_2 (ppm)	$\delta^{13}C$-CO_2 (‰)	Replicates	Root $\delta^{13}C$-CO_2	Soil $\delta^{13}C$-CO_2
Chaparral	1	400	−8.00	1	−26.5	−26.7
	2	430	−9.33	2	−27.4	−26.9
	3	450	−10.11	3	−30.2	−25.6
	4	480	−11.17	4	−25.8	−25.7
	5	500	−11.80			
	6	520	−12.38			
	7	550	−13.18			
	8	600	−14.33			
Pasture	1	390	−7.80	1	−27.7	−24.6
	2	435	−9.90	2	−29.9	−26.2
	3	450	−10.51	3	−28.6	−26.3
	4	482	−11.67	4	−31.2	−26.7
	5	500	−12.27			
	6	530	−13.16			
	7	549	−13.68			
	8	605	−15.01			
Woodland on a former corn field	1	402	−8.20	1	−28	−18.2
	2	428	−9.12	2	−26	−17.8
	3	455	−9.97	3	−28.6	−16
	4	478	−10.62	4	−28.6	−17
	5	507	−11.35			
	6	521	−11.67			
	7	545	−12.19			
	8	610	−13.38			

First, calculate the $\delta^{13}C$-CO_2 of soil respiration by applying the Keeling plot method (see Section 4.5). Then, estimate the relative contribution of roots (autotrophic) and soil (heterotrophic) to the total soil respiration flux, for the ecosystems for which this is possible (see Section 4.7). Last, calculate the standard error of your respiration estimates for each ecosystem.

Exercise 2

Examine diet sources in isopods

Let's imagine you are studying the effects of an invasive plant species on a population of isopods that feed exclusively on plant litter. Your study site is a C3–C4 grassland, but recently a new C3 N-fixing plant has invaded the site, and its cover is rapidly increasing. You want to know whether the isopods feed on the invasive plant and, if so, the percentage of their diet that is derived from it. You decide to use stable isotopes to answer your research question.

Next, you collect the following plant litter samples:

- C3 native grasses.
- C4 native grasses.
- C3 N-fixing invasive plant.
- Mixed sample of litter from which to extract the isopods.

After you extract the isopods from the mixed plant litter, you oven dry and pulverize all your samples. Then, you analyze them on an elemental analyzer (EA)-IRMS for C%, N%, $\delta^{13}C$, and $\delta^{15}N$. The results are reported in Table A4.2. You did not have the resources to quantify the actual isotope fractionation during isopods nutrition, which would have been ideal. Instead, you reviewed the literature on isotope fractionation in animal nutrition and decided to apply the average discrimination factors for invertebrates of $\Delta = 0.25‰$ for ^{13}C, and a $\Delta = 2.5‰$ for ^{15}N determined by Caut et al. (2009).

Referring to Sections 4.3 and 4.6, and using the data in Table A4.2, estimate the relative contribution of the three litter sources to the isopods' diet. Do the isopods feed on the invasive plant litter? If so, what percentage of the isopods' diet is derived from the invasive plant litter?

Table A4.2 *Results of laboratory analysis on plant litter and isopod samples. Values are realistic but made up for the purpose of this exercise.*

Component	$\delta^{13}C$ (‰)	C%	$\delta^{15}N$ (‰)	N%
Isopods	−19.2	NA	6.2	NA
C3 native grass	−25.6	45	6	0.2
C4 native grass	−13.2	45	5.5	0.4
C3 invasive plant	−27.1	45	0.3	2

Review questions

1. In Exercise 1, we suggested that you apply the Keeling plot approach, since you had measurements for both CO_2 concentrations and $\delta^{13}C$-CO_2 at eight points in time. Could you have applied alternative methods? Describe any alternative methods to calculate the $\delta^{13}C$-CO_2 of soil respiration. Explain the pros and cons of each with respect to the Keeling plot approach.

2. What criteria need to be met to successfully apply an isotope mixing model in a natural abundance scenario? Were you able to partition soil respiration for all the three ecosystems in Exercise 1?
3. In Exercise 2, what did you have to consider to be able to calculate the proportion of food sources for the isopods, and why?

References

Caut, S., E. Angulo and F. Courchamp (2009). "Variation in discrimination factors ($\Delta^{15}N$ and $\Delta^{13}C$): the effect of diet isotopic values and applications for diet reconstruction." *Journal of Applied Ecology* **46**(2): 443–453.

Phillips, D. L. and J. W. Gregg (2001). "Uncertainty in source partitioning using stable isotopes." *Oecologia* **127**(2): 171–179.

5

Heavy Isotope Enrichments

5.1 Isotope labeling

While the variation in natural abundance of stable isotopes generated by fractionation processes may offer the opportunity to trace the elemental contribution of a pool to another, often that variation is not large enough for accurate tracing. In these cases, isotope labeling comes in handy. Isotope labeling consists of the manipulation of the heavy isotope abundance (e.g., ^{13}C, ^{15}N, ^{18}O, or ^{2}H), generally by increasing it through additions of small amounts of the elemental pool in question with a higher proportion of the heavy isotope (e.g., $^{13}CO_2$, $^{15}NH_4$, $^{2}H_2O$, or $H_2{}^{18}O$). Isotope labeling experiments aim at modifying the heavy isotope abundance, without altering the system. Thus, particular care needs to be applied to avoid any other environmental alterations (e.g., changing pool size, or climate conditions). When isotope labeling generates large increases of the heavy isotope abundance, natural fractionation processes can be neglected. As mentioned in Chapter 2, at high levels of heavy isotope enrichment (>10%), linearity between the δ and the atom % notation is lost, and the δ notation becomes inaccurate. Thus, atom % is the recommended notation in isotope labeling studies with heavy isotope enrichments over 10 atom %.

Depending on the research questions and isotope of interest, different isotope-labeling techniques can be adopted. Here, we describe the most common methods used to label plants or other components of terrestrial ecosystems with ^{13}C, ^{15}N, and heavy water isotopes. Any labeling study requires a good knowledge of the study system, and a good model to interpret the data (Fry 2006). In particular, if labeling is used to trace the fate of an element in the environment, or determine process rates, fundamental knowledge of the exchange processes of that element through different molecules is required. For example, O and H isotopes quickly exchange between molecules (e.g., CO_2 and H_2O; ammonia ($NH_4{}^+$) or N_xO_x and H_2O), thus they do not accurately trace C or N compounds in the environment (e.g., Kool et al. 2009). Another problem to be aware of when planning isotope labeling experiments is the fact that some molecules (e.g., $NH_4{}^+$) may adsorb onto surfaces, reducing isotope recovery, and compromising experimental success. Heavy isotope labeling of natural systems will remain as a legacy in the system, impeding future on-site natural abundance studies. Similarly, isotope enrichment may contaminate laboratory space if not performed with care and followed by accurate cleaning after processing of isotope-enriched samples.

A Primer on Stable Isotopes in Ecology. M. Francesca Cotrufo and Yamina Pressler, Oxford University Press.
 DOI: 10.1093/oso/9780198854494.003.0005

These considerations can help address the first question that comes up in every isotope enrichment study: by how much should the heavy isotope be enriched? Isotope labeling is expensive and contaminates the natural isotopic environment. Thus, our advice is to keep the heavy isotope enrichment at the minimum level that enables measuring the isotopic tracer in the pools of interest. This level may vary from 99 atom % to a few atom % enrichment. You'll need to do your back-of-the-envelope calculation to find that out (see Chapter 5 activity)! A few things to consider: uncertainty in source partitioning decreases with the increasing isotopic differences between end members (Phillips and Gregg 2001), and fractionation can be neglected above an isotopic enrichment of about 500‰ (Fry 2006). Isotope labeling is a very exciting and powerful tool, which needs to be used with both knowledge and caution.

5.2 Carbon-13 labeling of plants: continuous versus pulse labeling

Carbon enters terrestrial ecosystems through plant photosynthesis, and because of this, the CO_2 pool is an ideal place to start when labeling plants and their C inputs with isotopic enrichment. To do so, the CO_2 pool that the plants have access to is enriched with ^{13}C. This can be done in the field or in a greenhouse, and in continuous or pulse mode. Continuous labeling refers to the continuous growth of plants from seeds or small seedlings to maturity under ^{13}C-CO_2 enriched atmospheres. Pulse labeling is instead achieved by periodic short-term exposures (hours to days) of plants to atmospheres enriched in ^{13}C-CO_2.

Several ^{13}C labeling chambers have been described in the literature, both using a continuous-labeling (e.g., Soong et al. 2014; Figure 5.1) or a pulse-labeling approach (e.g., Ostle et al. 2000; Bromand et al. 2001; Bird et al. 2003). In either case, chambers need to be transparent to photosynthetic radiation, and sealed as best as possible to prevent losses of the costly isotope label. CO_2 can be introduced directly as a gas at the desired ^{13}C enrichment level or generated from the acidification of carbonates. Since ^{13}C-enriched carbonates are generally cheaper and easier to acquire, the latter design has been used more often. In countries where cylinders of ^{13}C-enriched CO_2 are not readily available, the enriched carbonate approach is the only solution for CO_2 enrichment (Mitchell et al. 2016). Temperature, humidity, and CO_2 concentrations should also be monitored during labeling, and kept at the ambient level to avoid unintentionally altering photosynthetic rates through modifications of the environment. If labeling chambers are built within greenhouses, a reliable cooling system is also required.

CO_2 labeling was initially developed by using the radioactive ^{14}C isotope. Meharg (1994) offers a good review of the pros and cons of different ^{14}C labeling techniques used to trace C flows between plants and soils. These pros and cons equally apply to ^{13}C pulse- and continuous-labeling approaches, and we summarize them in Table 5.1. One should consider the following critical points when deciding on a CO_2 labeling method: the homogeneous distribution of ^{13}C in the labelled plant, the ability to close a ^{13}C budget within the system, the ability to trace the ^{13}C through time, and the cost and feasibility of the experiment.

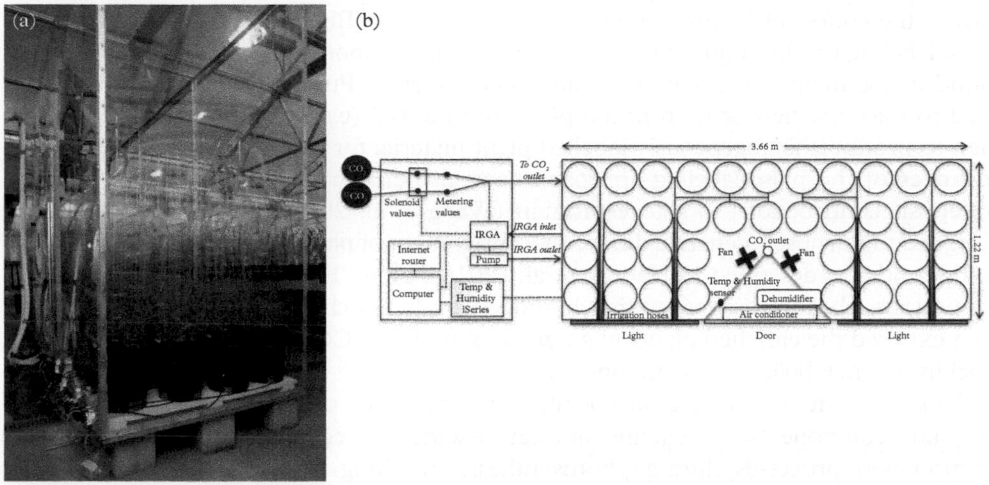

Figure 5.1 *Photo (a) and schematic (b) of continuous* ^{13}C *and* ^{15}N *stable isotope labelling chamber at Colorado State University. Schematic shows chamber from a bird's eye view. Broken lines represent electrical wiring. Solid lines represent gas or irrigation tubing.*
Modified from Soong et al. (2014).

Table 5.1 *Comparison of continuous and pulse* ^{13}C *isotope labeling approaches.*

Continuous labeling	**Pulse labeling**
Material is homogenously labeled, especially if plants are started from seeds	Material is not homogenously labeled. Labeling occurs in predominately labile C pools
Allows for complete C budgeting	Allows for tracing of recently assimilated C fluxes
Does not allow for discrimination between root exudation and root turnover	Label will largely manifest as root exudates
Poor time resolution	Good time resolution
High isotope enrichment of soil biota and organic matter pools	Low isotope enrichment of soil biota and organic matter pools
Cumbersome, expensive, and difficult to perform in the field	Versatile and applicable in the field
Quantitative assessment is possible	Mostly qualitative assessment is possible

As reviewed in Meharg (1994).

A homogeneous distribution of the ^{13}C label across the plant components is best achieved if plants are grown from seeds or small seedlings in a continuous-labeling chamber (Soong et al. 2014). However, repeated exposure of the plant to the label during the course of its development may also suffice (Bird et al. 2003). Occasional pulse labeling results in uneven enrichment of plant components, with most of the label found in the metabolic components and root exudates. Pulse labeling should thus be used to trace the flow of C from the plant into the soil (e.g., Ostle et al. 2000) rather than to produce homogeneously labelled plant material for subsequent studies. Recent advances of the pulse-labeling approach have also enabled the quantification of rhizodeposit contributions to soil respiration (Wang et al. 2021). Labeling can also be designed to purposely generate differential enrichment of plant structural and metabolic compounds, as described in Haddix et al. (2016). Here, Haddix et al. (2016) used a continuous-labeling chamber to first homogeneously enrich all plant components, and then exposed the enriched plants to a natural abundance CO_2 atmosphere to dilute the label in the metabolic plant components.

While continuous labeling will result in the presence of the heavy isotope across all plant components, it remains unclear whether it reduces or amplifies natural fractionation processes during photosynthesis, resulting in plant compounds with smaller (reduces) or greater (amplifies) isotopic differences than at natural abundance. Compound-specific analyses of ^{13}C-labeled plant materials are required to identify if any such patterns exist.

5.3 Amendments of isotope-labeled substrates and stable isotope probing

After photosynthesis, C circulates through terrestrial ecosystems in a myriad of different chemical forms. Many such C compounds can be purchased at the desired level and position of ^{13}C enrichment. Companies produce ^{13}C-enriched tissues from plants or other organisms (e.g., IsoLife), or you can produce your own using labeling chambers (Figure 5.1) or growth cultures. ^{13}C-enriched plant or microbial tissues, as well as specific C compounds can be added to the system of interest in the laboratory or the field. This approach is appropriate for research questions that involve quantifying flux rates or turnover times of specific C pools, tracing the fate of C within the soil food web and other organic or inorganic C pools, identifying the organisms involved in specific C metabolism, or quantifying C use efficiencies. Many such studies have been conducted and have been fundamental in advancing our understanding of C cycling in terrestrial ecosystems, from the study of priming (Fontaine et al. 2007), to the formation and persistence of soil organic matter (Cotrufo et al. 2015), or the fate of low-molecular-weight C compounds (Strickland et al. 2012), just to mention a few.

Similarly to C, substrates with different levels of enrichment of other stable isotopes, such as ^{15}N, ^{2}H, or ^{18}O, can be purchased or produced. Of particular interest are studies which combine the use of two isotope enrichments (e.g., ^{13}C and ^{15}N) because they also enable source partitioning of three pools (Haddix et al. 2016; see Section 4.6).

One of the most fascinating uses of stable isotope-enriched substrates is *stable isotope probing* (SIP), which consists of the labeling of microbial biomarkers to directly link active microbial groups to ecosystem functions (Boschker and Middelburg 2002)

(Figure 5.2). SIP was first developed by Radajewski et al. (2000), who demonstrated how ^{13}C-DNA produced by the group of microorganisms which fed on ^{13}C-labeled methane can be isolated from ^{12}C-DNA by density gradient centrifugation. Now, SIP is used in a vast array of applications and it can be particularly powerful when combined with the continuously advancing microarrays and metagenomic technologies (Dumont and Murrell 2005). Useful guidelines for the use of DNA (Neufeld et al. 2007), or RNA (Manefield et al. 2002) as biomarkers in SIP are available. However, density separation

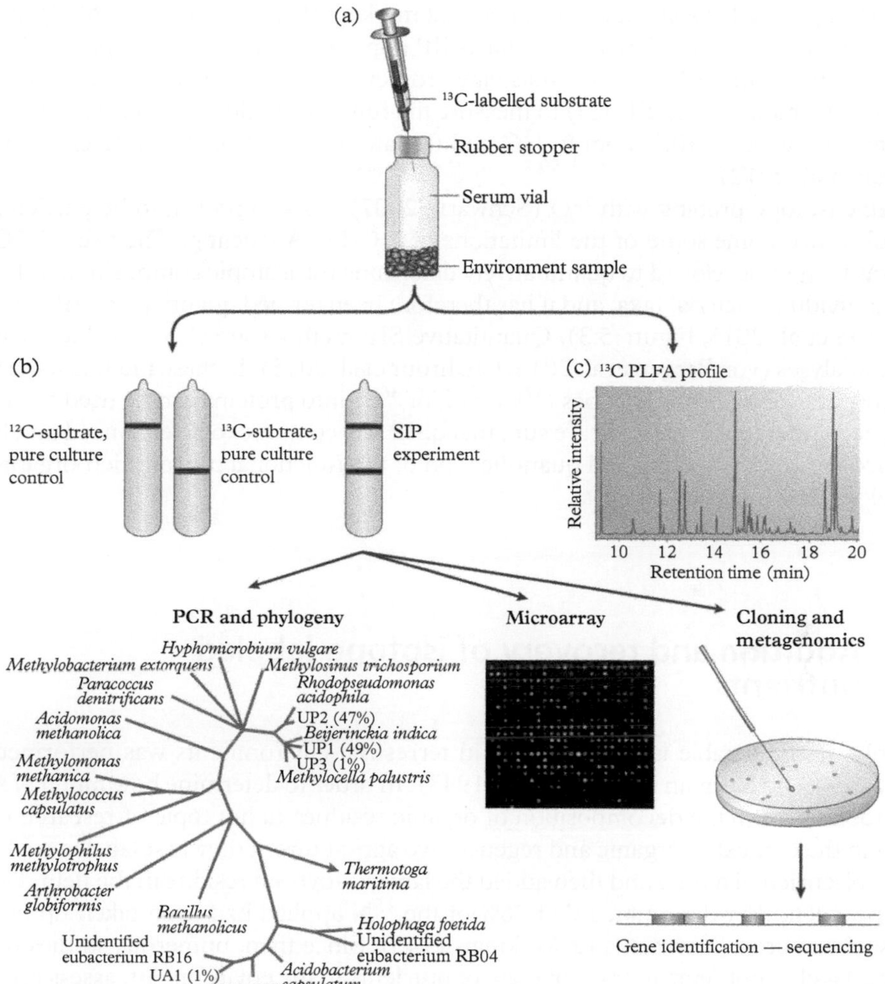

Figure 5.2 *Conceptual diagram of ^{13}C stable isotope probing (SIP) for DNA and phospholipid fatty acids (PLFAs) techniques.*

Reproduced with permission from Dumont and Murrell (2005).

of DNA or RNA requires high levels of enrichment to isolate an enriched DNA or RNA band. Thus, it is generally performed under high heavy isotope enrichment and controlled environments.

The use of phospholipid fatty acids (PLFAs) as biomarkers enables SIP at relatively low levels of isotope labeling (Rubino et al. 2010), and thus is an easier method to perform in the natural environment (Denef et al. 2009). However, PLFAs provide a very coarse resolution of the microbial community. Similarly, aminosugars can be used as biomarkers in SIP studies to look at substrate incorporation in fungal and bacterial tissues (Bodé et al. 2013), or at the permanence of microbial necromass in soil, since aminosugars are longer lived in soil than other microbial markers (Glaser and Gross 2005). More recently a phylogenetic microarray (Chip)-SIP approach has been developed and used in conjunction with secondary ion mass spectrometer imaging (i.e., nano secondary ion mass spectrometry (NanoSIMS)) to measure the functional role of uncultured microorganisms at low levels (0.5 atom % ^{13}C and 0.1 atom % ^{15}N) of substrate enrichment (Mayali et al. 2012).

Stable isotope probing with ^{18}O (Schwartz 2007) has also proven to be particularly helpful to overcome some of the limitations of ^{13}C-DNA labeling. The use of ^{18}O in SIP was further developed to quantitatively determine the isotopic composition of DNA from individual bacterial taxa, and it has therefore been termed quantitative SIP, or qSIP (Hungate et al. 2015; Figure 5.3). Quantitative SIP methods are also available for proteome analyses (von Bergen et al. 2013; Chahrour et al. 2015). In these methods, the rate of incorporation of heavy isotopes (^{13}C, ^{15}N, or ^{36}S) into proteins can be used to quantify specific metabolic rates. We are sure that qSIP will continue to be enhanced and help advance our understanding and quantification of the functional role of microorganisms in ecology and biogeochemistry.

5.4 Addition and recovery of isotope-labeled nutrients

Possibly the first stable isotope labeling in terrestrial environments was performed as early as 1943 by Norman and Werkman (1943). In order to determine how much N soybean took up from the decomposition of organic residues (a hot topic of research even today, in the context of organic and regenerative agriculture!), they first labeled soybean with ^{15}N-enriched nitrate and then added the labeled soybean residue in the field. Using this approach, they determined that 26% of the ^{15}N applied had been taken up by the new soybean crop (Norman and Werkman 1943). Since then, numerous studies using isotope-labeled nutrients to trace the fate of nutrients in the environment, assess rates of transformations, and determine fertilizer use efficiencies have been performed (Schimel 1993). Most commonly, the isotope-labeled nutrient is added in solution within the growth media, but some studies have also sprayed the labeled nutrient solution directly on the leaves (Zeller et al. 1998), added the labeled nutrient as a dry powder (Willison et al. 1998), or added the label as a gas (Murphy et al. 1998).

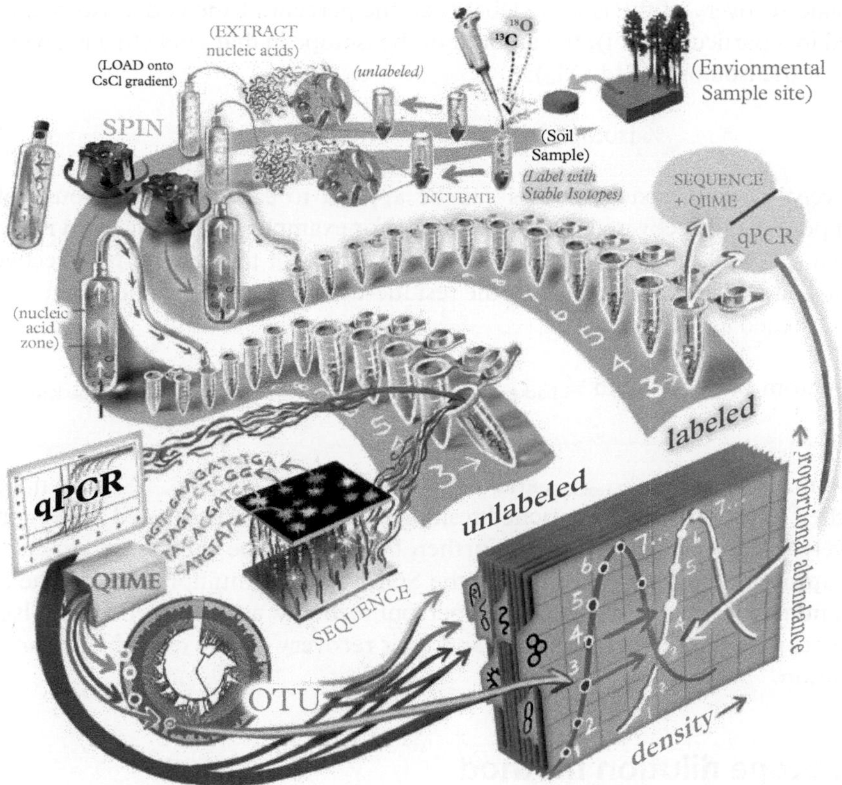

Figure 5.3 *Conceptual diagram of quantitative stable isotope probing (qSIP) approach, showing each step in the technique from collecting an environmental sample to determining the density of 16S rRNA gene fractions for individual taxa and their associated isotopic composition in atom %. Artwork is by Victor Leshyk.*

From Hungate et al. (2015). Reproduced with permission from Victor O. Leshyk, Center for Ecosystem Science and Society, Northern Arizona University.

Isotope tracing studies aim to calculate labeled isotope recovery in any target pool of interest. These studies commonly use the atom percent excess (atom % excess or APE) as the heavy atom fraction in a sample minus the natural heavy isotope abundance (background) to calculate the amount of the isotope label that is in a sample pool following isotope addition.

$$\text{Atom \% excess (APE)} = \text{atom \%}_{\text{enriched pool}} - \text{atom \%}_{\text{background pool}} \quad (5.1)$$

Thus, to calculate the mass of the label in a target pool (M_l), the APE is multiplied by the pool size for that element (M_e) (Goodale et al. 2015):

$$M_l = APE \cdot M_e \quad (5.2)$$

To calculate recovery of the isotope label (i.e., the percent of the added isotope that is recovered in a particular pool), the amount of the isotope in that pool (M_l) is divided by the mass of the isotope added (M_a):

$$\% \text{ isotope label recovery} = M_l/M_a \cdot 100 \quad (5.3)$$

Alternatively, isotope mixing models can be applied to calculate the amount of the enriched pool that is recovered in another pool. For example, if a ^{15}N-labeled residue is added to a soil, and the goal is to determine how much of the residue N is recovered, let's say in soil organic matter (SOM), the residue-derived fraction of the SOM ($f_{residue}$) can be calculated as:

$$f_{residue} = (\text{atom } \%_{SOM} - \text{atom } \%_{background\ SOM}) / (\text{atom } \%_{residue} - \text{atom } \%_{background\ SOM}) \quad (5.4)$$

Where atom $\%_{SOM}$, atom $\%_{residue}$, and atom $\%_{background\ SOM}$ are the ^{15}N percentages in the enriched SOM sample, the enriched residue added, and the unlabeled SOM control, respectively (e.g., Soong et al. 2016). Further, the $f_{residue}$ value can be used to calculate the total amount of residue-derived N in the SOM pool, by multiplying it by the SOM N pool, and that can be used to calculate percent recovery, as illustrated above. Isotope enrichments persist in the environment and their recovery can be quantified years after their addition.

5.5 Isotope dilution method

Isotope enrichment is also a powerful method to determine rates of transformations. To this end, the *isotope dilution method* is commonly used. First proposed by Kirkham and Bartholomew (1954, 1955) to determine rates of N mineralization, this method is now widely applied to determine gross rates of transformation of different nutrients in the environment (Di et al. 2000). The isotope dilution technique consists of labeling an inorganic nutrient pool (e.g., NH_4^+, NO_3^-, SO_4^{2-}) and measuring the changes in size and excess tracer abundance of the labeled nutrient pool over time. When nutrient pools are enriched with stable isotopes (e.g., ^{15}N, ^{34}S), the APE is commonly used to determine the excess tracer abundance. This technique has also been applied to determine P transformations by using radioactive P isotopes, in which case the specific activity of the pool is measured over time (Di et al. 1997).

The concept behind the isotope dilution method is that the tracer excess in a pool decreases over time proportionally to the rate of the input and output fluxes (Figure 5.4). This method assumes the following:

1. Isotope fractionation is negligible during the nutrient transformation process, because of the high isotope enrichment.
2. The influx nutrient has the background natural abundance isotopic composition (i.e., it is not labeled).

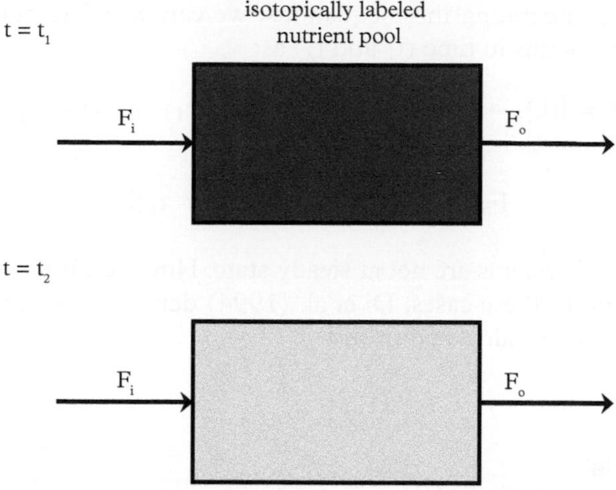

Figure 5.4 *Conceptual diagram of isotope pool dilution method. F_i represents influx and F_o represents outflux of a given nutrient. The tracer pool is diluted over time and the abundance of the tracer decreases, shown by the lighter color in the pool at $t = t_2$.*
Modified from Di et al. (2000).

3. At any point in time, the outflux nutrient has the same isotopic composition of the labeled nutrient pool at that point in time.
4. Both the influx and outflux rates are constant over the period of study.

Let's see how this works. Based on these assumptions, if we add and uniformly distribute an amount of label (q) to a nutrient pool, the label will be diluted over time because the influx (F_i) will contribute unlabeled nutrient to the pool. The outflux (F_o) will export the labeled and unlabeled nutrient out of the pool in the same proportion as they are in the pool at any point in time. If the system is not at steady state (i.e., the influx is different from the outflux), the rate (d_q/d_t) at which the added pool is lost through the outflux (F_o) becomes:

$$d_q/d_t = -F_o \cdot A\,(t) \tag{5.5}$$

where q is the quantity of the added tracer and A(t) is the tracer excess abundance (APE) at time t.

If Q is the total quantity of the nutrient pool, then its change over time is controlled by the nutrient influx (F_i) and outflux (F_o), as:

$$d_Q/d_t = F_i - F_o \tag{5.6}$$

The amount of the added tracer remaining in the nutrient pool at time t then becomes:

$$q\,(t) = Q\,(t) \cdot A\,(t) \tag{5.7}$$

By rearranging and integrating these equations, we can calculate both the influx and outflux between two points in time (t_1 and t_2) as:

$$F_i = [(Q_1 - Q_2) \cdot \ln(A_1/A_2)] / [(t_2 - t_1) \cdot \ln(Q_1/Q_2)] \tag{5.8}$$

$$F_o = F_i - [(Q_2 - Q_1) / (t_2 - t_1)] \tag{5.9}$$

In most cases, natural systems are not at steady state. However, if they are, $F_i = F_o$ and Q becomes constant. In these cases, Di et al. (1994) demonstrated that the flux rate F equal to F_i and F_o can be calculated from:

$$A(t) = A_0 \cdot e^{[-(F/Q)t]} \tag{5.10}$$

And, by rearranging:

$$F = [\ln(A_0/A_t)] \cdot Q/t = [\ln(A_1/A_2)] \cdot Q/(t_2 - t_1) \tag{5.11}$$

These numerical solutions (Di et al. 2000) of the isotope dilution method, as mentioned above, are based on the assumption that the flux rates do not change during the time of study. This may hold true over longer periods of time, but in short-term studies, when remineralization and nutrient immobilization processes may modify flux rates, analytical solutions are deemed more appropriate. Several have been proposed and Smith et al. (1994) provide a very useful comparison of the different analytical and numerical solutions to the isotope dilution method. Both approaches have strengths and weaknesses and before choosing the one to use, it is important to understand all their assumptions and consider which one best applies to the specific system in study.

The isotope dilution method has several limitations that are important to be aware of before applying it and when interpreting the results. Di et al. (2000) reviewed them extensively and we summarize the most important points here.

Multiple input and output fluxes. Most nutrient pools have multiple processes that contribute to the influx and outflux rates. Let's take, for example, mineral N fluxes (Figure 5.5). While NH_4^+ is produced by N mineralization and NO_3^- from nitrification, they can be lost through multiple processes in the natural environment where soil mineral N is lost through plant uptake, leaching and, in the case of ammonia, volatilization, and clay immobilization (Figure 5.5). When more than one process is involved, simplifying the system (e.g., by working in intact soil cores) and combining dilution studies so that the influx rate of one study is a contributor to the outflux of another study (i.e., nitrification in NO_3^- and NH_4^+ dilution, respectively) can help us interpret the measured fluxes (Davidson et al. 1991).

Label distribution. A basic assumption of the isotope dilution method is that the label is uniformly distributed within the pool of interest. When working with heterogeneous pools (e.g., soil), reaching complete uniformity is practically impossible. However, adding the label at multiple points (e.g., along the depth profile of interest) is highly recommended. When plants are present, labeling the rhizosphere and the bulk soil without

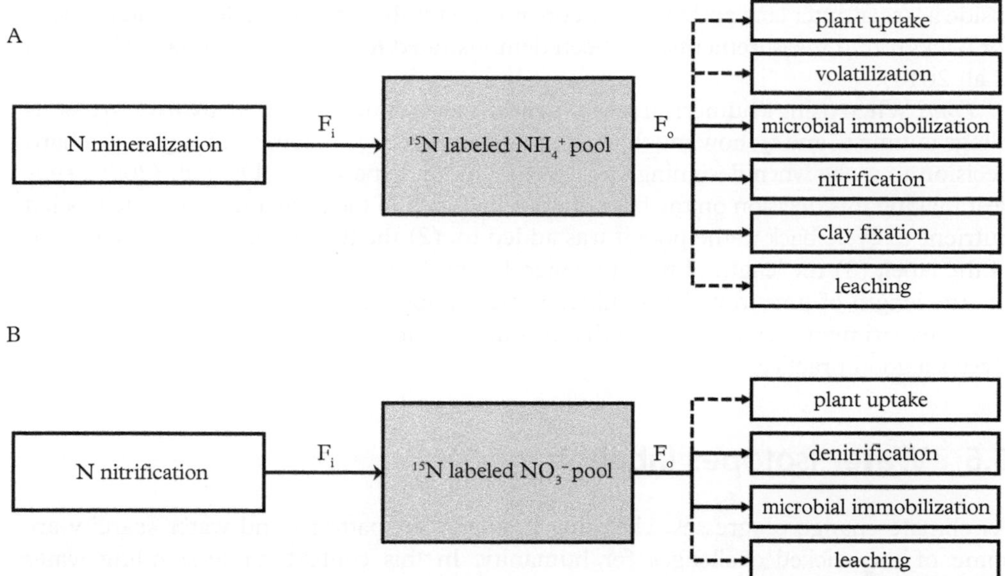

Figure 5.5 *Conceptual diagram for isotope pool dilution method to determine gross N transformation rates through the NH_4^+ pool (a) and the NO_3^- pool (b). Solid lines represent pathways that occur in laboratory and field conditions. Broken lines represent additional pathways that may occur during field studies.*
Modified from Di et al. (2000).

disrupting the soil environment is suggested. Non-uniform distribution of the label will cause significant errors in the F_i and F_o estimates, in particular when fluxes occur in hot spots. Such hot spots may occur when microorganisms are not evenly distributed across the labeled pool (Davidson et al. 1991), or when plants affect nutrient transformations more in the rhizosphere than bulk soil.

Environmental conditions. As for any other labeling experiment, the label addition should not affect the pool size or environmental conditions. This is particularly true in the case of the isotope dilution method, since here the goal is to accurately measure flux rates which are particularly sensitive to both pool size and environmental conditions. Scientists are used to finding the right compromise between alternative needs when designing experiments. This is one of those hard cases! For example, while multiple additions of the label are required to reach a uniform distribution, these may affect the pool environment by adding too much nutrient, too much water, or creating too many holes. Different solutions have been proposed, from adding tracers as dry powder (Willison et al. 1998), or as a gas (Murphy et al. 1998), to adding small label amounts at high enrichment levels (Di et al. 2000). Every experiment is different, and the researchers need to identify the most appropriate approach for their case. For soil mineralization studies in the field, using a two intact cores design, where a smaller inner core is placed inside a larger outer core and the inner core is used for the labeling while the outer is used for background measurements, has been demonstrated to be a wise approach (Koyama et al. 2010).

Time. When using a numerical solution, the isotope dilution method involves two samplings at time t_1 and t_2; how much time should pass between them? This again is a hard decision to make when designing an isotope dilution experiment. Di et al. (2000) suggest making this decision on the basis of four factors: (1) the time it takes for the labeled nutrient to cycle back to the pool it was added to, (2) the time it takes to lose detection of the label, (3) the length of time in which F_i and F_o can be considered constant, and (4) the length of time in which significant measurable nutrient transformations occur. Every experiment may require a different duration and running preliminary testing is always a good practice.

5.6 Water isotopes labeling

As climate change progresses, changing precipitation patterns and water scarcity are some of the wicked challenges for humanity. In this context, understanding water dynamics in terrestrial ecosystems is critical. Often, natural abundance isotopic differences between water sources are large enough to allow partitioning (e.g., Flanagan et al. 1992), but when that is not the case, water isotopes labeling becomes the solution.

Heavy isotope enrichment of water can be realized by labeling the water pool with either deuterium (^2H) or heavy oxygen (^{18}O or ^{17}O), and the choice most likely depends on the instrument available for monitoring the enriched water pool. The use of laser spectroscopy, as we will describe in Chapter 6, has made a breakthrough in the analysis of water isotopologues, facilitating the application of heavy water enrichments.

Here, we provide a few examples of water isotope enrichment applications. ^2H_2O water has been successfully applied to study hydraulic lifting (Peñuelas and Filella 2003) and deep root water uptake (Kulmatiski et al. 2010) by trees. In these studies, the ^2H_2O was either directly fed in vials to deep roots reaching an underground cave (Figure 5.6) (Peñuelas and Filella 2003), or added to the soil through deep injections (Kulmatiski et al. 2010). The former is a very creative method to trace water uptake by deep roots but is hard to repeat in most ecosystems. While adding enriched water through deep injections is a more repeatable approach, it requires significant care in avoiding large increases in the soil water pool, and thus altering the environmental conditions. Plant transpiration dynamics have been studied by injecting ^2H_2O straight in the transpiration stream of trees or bamboos by drilling holes at the base of the tree stem, or the bamboo culm (James et al. 2003; Dierick et al. 2010).

Water labeling experiments have also been conducted within controlled labeling chambers. Using a custom-made chamber with separate root and shoot compartments, Barthel and collaborators (2014) enriched the irrigation water of beech saplings to a $\delta^{18}O$ of 449 ± 7‰ in order to investigate the effect of drought on below- and aboveground fluxes, including on CO_2 released by leaves and soils. As discussed in Section 5.3, ^{18}O labeling of water has now become a very useful method to quantitatively label the DNA of active microbes (Hungate et al. 2015). Water labeling can also be realized

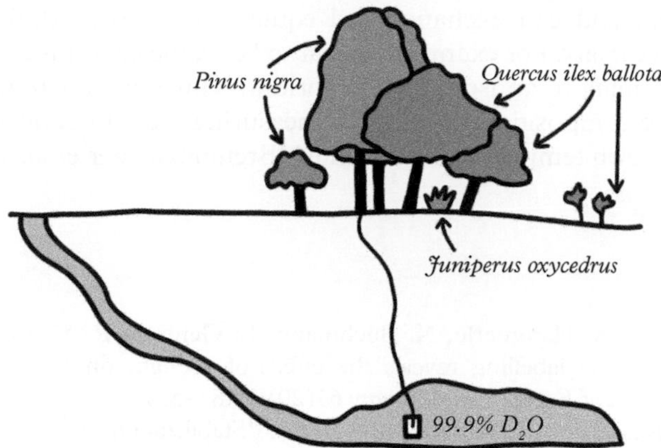

Figure 5.6 *Example of a creative approach to heavy isotope labeling where D_2O (^2H_2O) was directly fed to deep roots of* Pinus nigra *through an underground cave. The isotope signal was then traced throughout the plant community.*
Figure modified from Peñuelas and Filella (2003).

by depleting ^2H and ^{18}O abundance of the water pool. This is the approach used in the multi-isotope labeling chamber at the University of Zurich to produce organic matter depleted in heavy water isotopes. For this purpose, water vapor depleted in ^{18}O and ^2H ($\delta^{18}O = -370$‰ and $\delta_2H = -813$‰) was added continuously to plant shoots for 14 days (Studer et al. 2015).

Besides water enrichments, ^{18}O labeling has been used in an array of different applications. The first $^{18}O_2$ labeling experiment was conducted by Berry et al. (1978) using a thermostated leaf cuvette to study O_2 fixation during photosynthesis. This $^{18}O_2$ approach has since been expanded to study the rate of O_2 incorporation into specific compounds (e.g., glycolate, glycine, and serine) and to measure photorespiration (Jolivet-Tournier and Gerster 1984; de Veau and Burris 1989). Typically, highly enriched $^{18}O_2$ (e.g., 10–99 atom %) is added in an air mixture to the leaf cuvette at a fixed rate for a short period of time (e.g., seconds to minutes), before the leaf is harvested, plunged in liquid N_2, and analyzed for the ^{18}O enrichment of bulk tissues or compounds of interest. Gas measurements may be taken during the labeling, depending on the experimental objectives. The use of ^{18}O labeling has also been attempted to partition N_2O emission sources from soil (Kool et al. 2007). However, the same authors realized that the method was flawed because O atoms exchange between nitrogen oxides and water in soil (Kool et al. 2009).

Ozone (O_3) labeling has been used to trace the fate of ozone in plant–soil systems. The $^{18}O_3$ was generated from 95% atom enriched O_2 and added (O_3 concentration <5 nl/L) to a controlled-environment chamber ventilated by charcoal-filtered air (Toet et al. 2009).

Hydrogen and oxygen stable isotope labeling, while relatively easy to conduct, need to be used with a full understanding of the chemistry of the system. They are very mobile in the environment and can exchange and equilibrate between different molecules, compromising traceability. For example, similar to N_2O, the O atoms exchange between water and CO_2, coming to an isotopic equilibrium. This phenomenon is used to estimate the O isotope composition in water by measuring the ^{18}O concentration in CO_2 equilibrated at a known temperature with water (Brenninkmeijer et al. 1983).

References

Barthel, M., P. Sturm, A. Hammerle, N. Buchmann, L. Gentsch, R. Siegwolf and A. Knohl (2014). "Soil H2 18O labelling reveals the effect of drought on C 18OO fluxes to the atmosphere." *Journal of Experimental Botany* **65**(20): 5783–5793.

Bird, J. A., C. van Kessel and W. R. Horwath (2003). "Stabilization of ^{13}C-carbon and immobilization of ^{15}N-nitrogen from rice straw in humic fractions." *Soil Science Society of America Journal* **67**(3): 806–816.

Bodé, S., R. Fancy and P. Boeckx (2013). "Stable isotope probing of amino sugars—a promising tool to assess microbial interactions in soils." *Rapid Communications in Mass Spectrometry* **27**(12): 1367–1379.

Boschker, H. T. S. and J. J. Middelburg (2002). "Stable isotopes and biomarkers in microbial ecology." *Fems Microbiology Ecology* **40**(2): 85–95.

Brenninkmeijer, C. A. M., P. Kraft and W. G. Mook (1983). "Oxygen isotope fractionation between CO_2 and H_2O." *Chemical Geology* **41**: 181–190.

Bromand, S., J. K. Whalen, H. H. Janzen, J. K. Schjoerring and B. H. Ellert (2001). "A pulse-labelling method to generate ^{13}C-enriched plant materials." *Plant and Soil* **235**(2): 253–257.

Chahrour, O., D. Cobice and J. Malone (2015). "Stable isotope labelling methods in mass spectrometry-based quantitative proteomics." *Journal of Pharmaceutical and Biomedical Analysis* **113**: 2–20.

Cotrufo, M. F., J. L. Soong, A. J. Horton, E. E. Campbell, M. L. Haddix, D. H. Wall and W. J. Parton (2015). "Formation of soil organic matter via biochemical and physical pathways of litter mass loss." *Nature Geoscience* **8**(10): 776–779.

Davidson, E. A., S. C. Hart, C. A. Shanks and M. K. Firestone (1991). "Measuring gross nitrogen mineralization, and nitrification by ^{15}N isotopic pool dilution in intact soil cores." *Journal of Soil Science* **42**(3): 335–349.

de Veau, E. J. and J. E. Burris (1989). "Photorespiratory rates in wheat and maize as determined by ^{18}O-labeling." *Plant Physiology* **90**(2): 500.

Denef, K., D. Roobroeck, M. C. W. Manimel Wadu, P. Lootens and P. Boeckx (2009). "Microbial community composition and rhizodeposit-carbon assimilation in differently managed temperate grassland soils." *Soil Biology and Biochemistry* **41**(1): 144–153.

Di, H. J., K. C. Cameron and R. G. McLaren (2000). "Isotopic dilution methods to determine the gross transformation rates of nitrogen, phosphorus, and sulfur in soil: a review of the theory, methodologies, and limitations." *Soil Research* **38**(1): 213–230.

Di, H. J., L. M. Condron and E. Frossard (1997). "Isotope techniques to study phosphorus cycling in agricultural and forest soils: a review." *Biology and Fertility of Soils* **24**(1): 1–12.

Di, H. J., R. Harrison and A. S. Campbell (1994). "Assessment of methods for studying the dissolution of phosphate fertilisers of differing solubility in soil. I. An isotopic method." *Fertiliser Research* 38(1): 11–18.

Dierick, D., D. Hölscher and L. Schwendenmann (2010). "Water use characteristics of a bamboo species (Bambusa blumeana) in the Philippines." *Agricultural and Forest Meteorology* **150**(12): 1568–1578.

Dumont, M. G. and J. C. Murrell (2005). "Stable isotope probing—linking microbial identity to function." *Nature Reviews Microbiology* **3**(6): 499–504.

Flanagan, L. B., J. R. Ehleringer and J. D. Marshall (1992). "Differential uptake of summer precipitation among co-occurring trees and shrubs in a pinyon-juniper woodland." *Plant, Cell & Environment* **15**(7): 831–836.

Fontaine, S., S. Barot, P. Barré, N. Bdioui, B. Mary and C. Rumpel (2007). "Stability of organic carbon in deep soil layers controlled by fresh carbon supply." *Nature* **450**(7167): 277–280.

Fry, B. (2006). *Stable isotope ecology*. New York, Springer.

Glaser, B. and S. Gross (2005). "Compound-specific $\delta^{13}C$ analysis of individual amino sugars—a tool to quantify timing and amount of soil microbial residue stabilization." *Rapid Communications in Mass Spectrometry* **19**(11): 1409–1416.

Goodale, C. L., G. Fredriksen, M. S. Weiss, C. K. McCalley, J. P. Sparks and S. A. Thomas (2015). "Soil processes drive seasonal variation in retention of ^{15}N tracers in a deciduous forest catchment." *Ecology* **96**(10): 2653–2668.

Haddix, M. L., E. A. Paul and M. F. Cotrufo (2016). "Dual, differential isotope labeling shows the preferential movement of labile plant constituents into mineral-bonded soil organic matter." *Global Change Biology* **22**(6): 2301–2312.

Hungate, B. A., R. L. Mau, E. Schwartz, J. G. Caporaso, P. Dijkstra, N. van Gestel, B. J. Koch, C. M. Liu, T. A. McHugh, J. C. Marks, E. M. Morrissey and L. B. Price (2015). "Quantitative microbial ecology through stable isotope probing." *Applied and Environmental Microbiology* **81**(21): 7570.

James, S. A., F. C. Meinzer, G. Goldstein, D. Woodruff, T. Jones, T. Restom, M. Mejia, M. Clearwater and P. Campanello (2003). "Axial and radial water transport and internal water storage in tropical forest canopy trees." *Oecologia* **134**(1): 37–45.

Jolivet-Tournier, P. and R. Gerster (1984). "Incorporation of oxygen into glycolate, glycine, and serine during photorespiration in maize leaves." *Plant physiology* **74**(1): 108–111.

Kirkham, D. and W. V. Bartholomew (1954). "Equations for following nutrient transformations in soil, utilizing tracer data." *Soil Science Society of America Journal* **18**(1): 33–34.

Kirkham, D. and W. V. Bartholomew (1955). "Equations for following nutrient transformations in soil, utilizing tracer data: II." *Soil Science Society of America Journal* **19**(2): 189–192.

Kool, D. M., N. Wrage, O. Oenema, J. Dolfing and J. W. Van Groenigen (2007). "Oxygen exchange between (de)nitrification intermediates and H_2O and its implications for source determination of NO and N_2O: a review." *Rapid Communications in Mass Spectrometry* **21**(22): 3569–3578.

Kool, D. M., N. Wrage, O. Oenema, D. Harris and J. W. Van Groenigen (2009). "The ^{18}O signature of biogenic nitrous oxide is determined by O exchange with water." *Rapid Communications in Mass Spectrometry* **23**(1): 104–108.

Koyama, A., K. L. Kavanagh and K. Stephan (2010). "Wildfire effects on soil gross nitrogen transformation rates in coniferous forests of central Idaho, USA." *Ecosystems* **13**(7): 1112–1126.

Kulmatiski, A., K. H. Beard, R. J. T. Verweij and E. C. February (2010). "A depth-controlled tracer technique measures vertical, horizontal and temporal patterns of water use by trees and grasses in a subtropical savanna." *New Phytologist* **188**(1): 199–209.

Manefield, M., A. S. Whiteley, R. I. Griffiths and M. J. Bailey (2002). "RNA stable isotope probing, a novel means of linking microbial community function to phylogeny." *Applied and Environmental Microbiology* **68**(11): 5367.

Mayali, X., P. K. Weber, E. L. Brodie, S. Mabery, P. D. Hoeprich and J. Pett-Ridge (2012). "High-throughput isotopic analysis of RNA microarrays to quantify microbial resource use." *The ISME Journal* **6**(6): 1210–1221.

Meharg, A. A. (1994). "A critical review of labelling techniques used to quantify rhizosphere carbon-flow." *Plant and Soil* **166**(1): 55–62.

Mitchell, E., C. Scheer, D. W. Rowlings, R. T. Conant, M. F. Cotrufo, L. van Delden and P. R. Grace (2016). "The influence of above-ground residue input and incorporation on GHG fluxes and stable SOM formation in a sandy soil." *Soil Biology and Biochemistry* **101**: 104–113.

Murphy, D. V., G. P. Sparling and I. R. P. Fillery (1998). "Stratification of microbial biomass C and N and gross N mineralisation with soil depth in two contrasting Western Australian agricultural soils." *Soil Research* **36**(1): 45–56.

Neufeld, J. D., J. Vohra, M. G. Dumont, T. Lueders, M. Manefield, M. W. Friedrich and J. C. Murrell (2007). "DNA stable-isotope probing." *Nature Protocols* **2**(4): 860–866.

Norman, A. G. and C. H. Werkman (1943). "The use of the nitrogen isotope N15 in determining nitrogen recovery from plant materials decomposing in soil." *Agronomy Journal* **35**(12): 1023–1025.

Ostle, N., P. Ineson, D. Benham and D. Sleep (2000). "Carbon assimilation and turnover in grassland vegetation using an in situ $^{13}CO_2$ pulse labelling system." *Rapid Communications in Mass Spectrometry* **14**(15): 1345–1350.

Peñuelas, J. and I. Filella (2003). "Deuterium labelling of roots provides evidence of deep water access and hydraulic lift by Pinus nigra in a Mediterranean forest of NE Spain." *Environmental and Experimental Botany* **49**(3): 201–208.

Phillips, D. L. and J. W. Gregg (2001). "Uncertainty in source partitioning using stable isotopes." *Oecologia* **127**(2): 171–179.

Radajewski, S., P. Ineson, N. R. Parekh and J. C. Murrell (2000). "Stable-isotope probing as a tool in microbial ecology." *Nature* **403**(6770): 646–649.

Rubino, M., J. A. J. Dungait, R. P. Evershed, T. Bertolini, P. De Angelis, A. D'Onofrio, A. Lagomarsino, C. Lubritto, A. Merola, F. Terrasi and M. F. Cotrufo (2010). "Carbon input belowground is the major C flux contributing to leaf litter mass loss: evidences from a ^{13}C labelled-leaf litter experiment." *Soil Biology and Biochemistry* **42**(7): 1009–1016.

Schimel, D. S. (1993). *Theory and application of tracers*. San Diego, CA, Academic Press.

Schwartz, E. (2007). "Characterization of growing microorganisms in soil by stable isotope probing with H218O." *Applied and Environmental Microbiology* **73**(8): 2541.

Smith, C. J., P. M. Chalk, D. M. Crawford and J. T. Wood (1994). "Estimating gross nitrogen mineralization and immobilization rates in anaerobic and aerobic soil suspensions." *Soil Science Society of America Journal* **58**(6): 1652–1660.

Soong, J. L., D. Reuss, C. Pinney, T. Boyack, M. L. Haddix, C. E. Stewart and M. F. Cotrufo (2014). "Design and operation of a continuous ^{13}C and ^{15}N labeling chamber for uniform or differential, metabolic and structural, plant isotope labeling." *Journal of Visualized Experiments: JoVE* **83**: e51117.

Soong, J. L., M. L. Vandegehuchte, A. J. Horton, U. N. Nielsen, K. Denef, E. A. Shaw, C. M. de Tomasel, W. Parton, D. H. Wall and M. F. Cotrufo (2016). "Soil microarthropods support ecosystem productivity and soil C accrual: evidence from a litter decomposition study in the tallgrass prairie." *Soil Biology and Biochemistry* **92**: 230–238.

Strickland, M. S., K. Wickings and M. A. Bradford (2012). "The fate of glucose, a low molecular weight compound of root exudates, in the belowground foodweb of forests and pastures." *Soil Biology and Biochemistry* **49**: 23–29.

Studer, M. S., R. T. W. Siegwolf, M. Leuenberger and S. Abiven (2015). "Multi-isotope labelling of organic matter by diffusion of $^{2}H/^{18}O$-H_2O vapour and ^{13}C-CO_2 into the leaves and its distribution within the plant." *Biogeosciences* **12**(6): 1865–1879.

Toet, S., J.-A. Subke, D. D'Haese, M. R. Ashmore, L. D. Emberson, Z. Crossman, R. P. Evershed, J. D. Barnes and P. Ineson (2009). "A new stable isotope approach identifies the fate of ozone in plant–soil systems." *New Phytologist* **182**(1): 85–90.

von Bergen, M., N. Jehmlich, M. Taubert, C. Vogt, F. Bastida, F.-A. Herbst, F. Schmidt, H.-H. Richnow and J. Seifert (2013). "Insights from quantitative metaproteomics and protein-stable isotope probing into microbial ecology." *The ISME Journal* 7(10): 1877–1885.

Wang, R., B., Bicharanloo, M.B. Shirvan, T.R. Cavagnaro, Y. Jiang, C. Keitel, and F.A. Dijkstra (2021). "A novel ^{13}C pulse-labelling method to quantify the contribution of rhizodeposits to soil respiration in a grassland exposed to drought and nitrogen addition." *New Phytologist* **230**: 857–866.

Willison, T. W., J. C. Baker, D. V. Murphy and K. T. Goulding (1998). "Comparison of a wet and dry ^{15}N isotopic dilution technique as a short-term nitrification assay." *Soil Biology & Biochemistry* **30**(5): 661–663.

Zeller, B., M. Colin-Belgrand, E. Dambrine, F. Martin and P. Bottner (2000). "Decomposition of ^{15}N-labelled beech litter and fate of nitrogen derived from litter in a beech forest." *Oecologia* **123**(4): 550–559.

Zeller, B., M. Colin-Belgrand, É. Dambrine and F. Martin (1998). "^{15}N partitioning and production of ^{15}N-labelled litter in beech trees following [^{15}N]urea spray." *Annals of Forest Science* **55**(3): 375–383.

Activity: Designing Heavy Isotope Enrichment in Biogeochemical Studies

Objectives

- Understand important considerations when designing an isotope enrichment study.
- Evaluate how to achieve desired levels of heavy isotope enrichment.

Tools and background

This activity can be completed entirely by hand with only a calculator, but we recommend using a spreadsheet or data processing software to help organize your ideas. Refer to the equations for calculating addition and recovery of labeled nutrients in Section 5.4 to help you complete the activities.

There is not a set amount of heavy isotope that should be added in labeling experiments. The level of enrichment in a given study should be determined based upon the specific system, research questions, and limitations of the study. Heavy isotope addition experiments require careful planning to ensure that the design of the experiment supports the question that is being asked. Review Section 5.1 for further discussion on initial consideration in labeling experiments.

Exercise 1

Calculate the target $^{13}C\text{-}CO_2$ enrichment in a labeling chamber

Imagine you are in charge of the continues labeling chamber described in Soong et al. (2014) and want to achieve a target $^{13}C\text{-}CO_2$ atmosphere of 4.5 atom % ^{13}C in the chamber. However, you could not find a CO_2 tank with that level of enrichment available for purchase. Instead, you bought one tank certified at 10 atom % ^{13}C, and one tank at natural abundance (1.1 atom % ^{13}C), with the intention of mixing the two to achieve the desired level of enrichment in the chamber.

Calculate the relative contribution of CO_2 flux from each of the two tanks to the chamber atmosphere required to achieve the target $^{13}C\text{-}CO_2$ of 4.5 atom % ^{13}C. Which model did you use in the calculation?

Exercise 2

Calculate optimal ^{15}N enrichment for tracing mineral N additions in field studies

Imagine you are a forest biogeochemist studying N dynamics in a pine forest after a clearcut event. It was observed that clearcutting may result in high N losses with negative cascading impacts on the environment. Therefore, you want to quantify the amount of mineral N pine seedlings take up after clearcutting to better understand the relative plant N uptake versus N losses. NH_4^+ is the prevailing form of mineral N in the soils at the site. So, you decide to label the soil around the pine seedlings with $^{15}N\text{-}NH_4^+$, and then measure the ^{15}N in the newly formed pine needles.

For your experiment, you established six replicated plots of 1 m^2 each with one pine seedling in the middle. Each plot will receive the same amount of $^{15}N\text{-}NH_4^+$. Note that in a real experiment you would also want to set replicated control plots which would receive unlabeled ammonia, but we will only focus on the enriched plots for the purpose of this exercise.

In designing your experiment, you ask yourself:

- How much ^{15}N-labeled NH_4^+ should I add to each plot?
- What level of ^{15}N enrichment should I apply?

The following exercise will guide you through answering these two questions. You need to determine the appropriate amount of $^{15}N\text{-}NH_4^+$ to add without altering the NH_4^+ pool size, and decide that an addition of not more than 5% would be adequate to your scope. You also decide to achieve a target $\delta^{15}N$ enrichment value of the total soil N pool of 500‰. In fact, you had found that this level of enrichment will be large enough so that kinetic fractionation effects can be ignored but would avoid measurement problems that arise at very high enrichment levels (and would not be too expensive!). Please note that these values are realistic but made up for the purposes of this activity.

Exercise 2a

Determine the amount of nitrogen (NH_4-N) to add

Using the NH_4-N stock information given in Table A5.1, determine the average amount of g $NH_4\text{-}N/m^2$ at the site. Then calculate the amount of NH_4-N that can be added without changing the pool size by more than 5%, based on the average NH_4-N stock value across the replicate plots.

Table A5.1 *Amount of nitrogen in ammonium (N-NH_4) in soils in field study. Most of the roots are in the topsoil, thus you focus your study to the top 15 cm*

Replicate	g $NH_4\text{-}N/m^2$ (0–15 cm)
1	52.16
2	74.82
3	36.35
4	64.28
5	46.84
6	42.14
Average	

Data are from pine forest soils in Colorado (courtesy of Dr. B. Avera).

Exercise 2b

Determine the level of ^{15}N enrichment of the added $N\text{-}NH_4^+$

You want to achieve a ^{15}N atom % of 0.54845 (equivalent to a $\delta^{15}N$ of 500‰) in the labeled pool (i.e., the soil NH_4^+ in our study). The NH_4^+ at the site has a natural abundance $\delta^{15}N$ of 1‰ (0.36667 atom % ^{15}N). Calculate the ^{15}N atom % value of the g NH_4-N you will need to add to each plot (calculated in Exercise 2a) to reach the target enrichment of the labeled pool. What model are you going to use for the calculation?

Review questions

1. Isotope addition studies require many considerations prior to implementing the approach to address a research question. Are there any other challenges to consider when designing an isotope addition study that weren't considered in the exercises above? How would you work through the decision-making process for each consideration?
2. Based on your calculations in the exercises above, would you get the same results if you had used isotopic values in the δ notation instead of the atom % notation? Would the same apply if we had said that target enrichment was 90 atom % heavy isotopes? When working with isotope enrichment make always sure to use a correct notation!

Reference

Soong, J. L., D. Reuss, C. Pinney, T. Boyack, M. L. Haddix, C. E. Stewart and M. F. Cotrufo (2014). "Design and operation of a continuous ^{13}C and ^{15}N labeling chamber for uniform or differential, metabolic and structural, plant isotope labeling." *Journal of Visualized Experiments: JoVE* 83: e51117.

6

Measuring Stable Isotopes

6.1 Principles of measuring stable isotopes

An isotopic substitution within a molecule changes the mass of the molecule when a heavy isotope substitutes for a lighter one (or vice versa). Additionally, the isotopic substitution modifies the vibrational energy of the bonds which connect the isotope to the molecule, and therefore the electromagnetic radiation absorption properties of the molecule. In the case of isotopologues, which are molecules that differ in their heavy isotope abundance, both mass and energy vary. While in the case of isotopomers, molecules that have the same heavy isotope abundance (i.e., mass) but the isotopic substitution is in a different position (e.g., $^{14}N^{15}N^{16}O$ versus $^{15}N^{14}N^{16}O$), only their electromagnetic radiation absorption properties differ.

It is thus not surprising that mass differences are used to separate and quantify the abundance of isotopologues by an instrument defined as an isotope ratio mass spectrometer (IRMS), while electromagnetic radiation absorption is used to detect and quantify both isotopologues and isotopomers, through laser spectroscopy . Nuclear magnetic resonance spectroscopy, generally known as NMR, also detects isotopes by observing the magnetic field around atomic nuclei. However, NMR informs us about the electronic structure of a molecule and its functional groups but cannot be used for isotope analysis and therefore we do not discuss it here.

In this chapter, we will first describe the principle of IRMS and laser spectroscopy for the quantification of all isotopocules and then will present the preparation methods and instruments required to enable isotope analysis of samples of ecological interest in gas, liquid, or solid states. We will focus on H, C, N, O, and S isotope analysis. IRMS can only analyze isotopes in pure gas molecules (e.g., H_2, N_2, O_2, CO_2, CO, SO_2), while laser spectroscopy can analyze them in gas mixtures. Thus, any solid and liquid samples, and gas mixtures in the case of IRMS, require significant sample preparation to get to the point where they can be analyzed for their isotopic composition. We will see how much advancement in isotope analysis has been made in the past few decades in the area of laser spectroscopy and sample preparation, which today enables us to detect the isotope composition of air samples in remote ecosystems, such as the arctic tussock tundra (Lynch et al. 2018), or in solid samples down to the nanoscale (Mueller et al. 2012). Recent advancement in isotope analysis, alongside many other atomic spectrometry methods, were reviewed by Butler et al. (2018). It's exciting to imagine what may come next!

A Primer on Stable Isotopes in Ecology. M. Francesca Cotrufo and Yamina Pressler, Oxford University Press.
 DOI: 10.1093/oso/9780198854494.003.0006

6.2 Isotope ratio mass spectrometry

Ionized nuclei of different masses accelerated within a magnetic field under a vacuum deviate from their trajectory in a curve proportional to their masses. This property of nuclei was discovered by Francis W. Aston, who invented the first mass spectrometer in 1919. In this apparatus, the ingenious use of electromagnetic focusing enabled him to use the very slight differences in mass of the neon isotopes to obtain their separation. He later discovered around 212 of the natural isotopes and won the Nobel Prize in Chemistry in 1922. Great progress was made in the following years, and in 1940 Alfred O. C. Nier published the first design of a mass spectrometer suitable for routine C and N stable isotope analyses (Nier 1940). But it was in 1950 that McKinney and collaborators designed the first mass spectrometer with a sensitivity high enough to allow the measurement of C and O isotopes in CO_2 and O_2 gases (McKinney et al. 1950). Since then, the basic design of isotope ratio mass spectrometers (IRMSs) has remained the same. Improvements such as the development of the continuous injection (conflow) system have since been made and further improvements regarding the attainable precision, dynamic sample size, and throughput continue to be made. Detailed descriptions of IRMS technology has been covered elsewhere (e.g., Voglar et al. 2019), and we present a brief overview here.

An IRMS is made of (1) an injection system which can be either a dual inlet or, now much more commonly, a conflow ; (2) an ion source; (3) a magnetic field; (4) Faraday cup detectors and amplifiers; (5) a pumping system; and (6) a computer (Figure 6.1).

Pure gas molecules (i.e., H_2, N_2, O_2, CO_2, CO, SO_2) are injected into an IRMS . Originally, a dual inlet system was used in IRMS to alternatively inject samples through one inlet and a standard through the other, thus assuring high precision. The later discovery of the conflow injection system allowed for higher sample injection throughput and to directly connect an IRMS to other preparation systems (e.g., elemental analyzers (EAs), gas chromatographs, etc.) which output the pure gas. However, conflow injection never reached the same precision of dual inlet, thus the latter is still the best approach for very high precision applications.

Once the gas molecules are injected into the ion source, they are ionized by an ion beam generated by a hot tungsten filament. The ionized molecules are then accelerated into a magnetic field. Because the strength of the magnetic field and the acceleration potential are held constant, ionized molecules' trajectories are a sole function of their mass and energy. Ion molecules of smaller masses are deflected more than ion molecules of higher masses. By positioning Faraday cups at the end of the trajectory of specific masses (e.g., 44, 45, and 46 for CO_2 isotopologues, or 2 and 3 for H and ^{2}H; Figure 6.1) ions can be counted based on the number of strikes on the specific cup collector. As ions strike on a Faraday cup collector they neutralize and generate an electrical current which is amplified and then used to calculate the isotope ratio and δ values by the computer software. The abundance of each isotopic mass is also reported by the computer allowing for manual calculation of ratios when needed. A pumping system maintains the high vacuum needed in the IRMS.

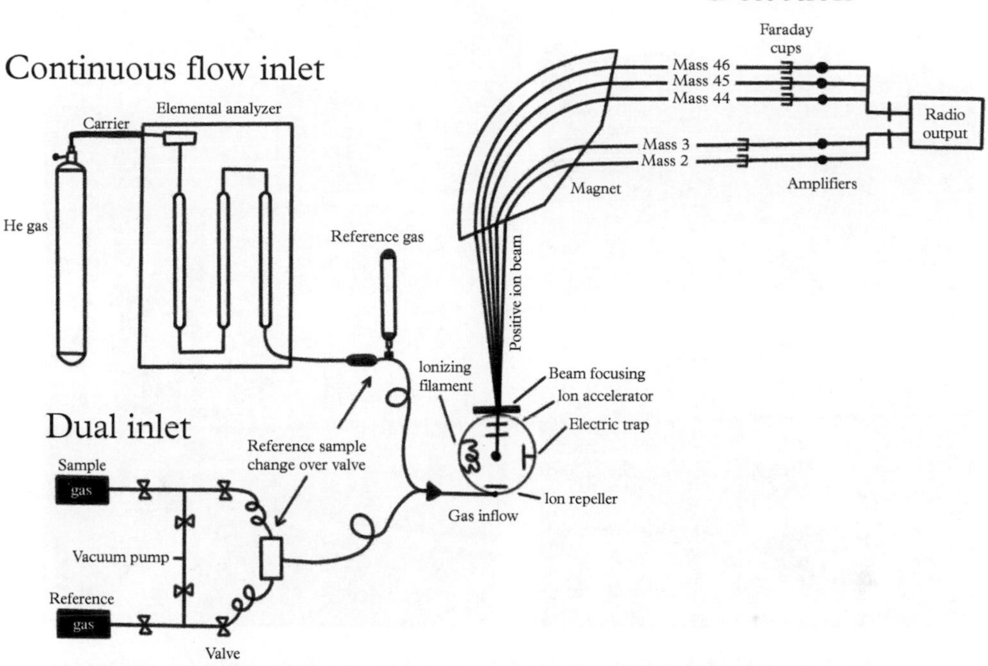

Figure 6.1 *General schematic of an isotope ratio mass spectrometer (IRMS).*
Modified from http://web.sahra.arizona.edu/programs/isotopes/methods/gas.html

Several IRMS instruments are on the market, with designs that continuously advance their precision, ability to reduce sample amount, measure more isotopes simultaneously, and increase intuitiveness of software design and operation. The IRMS remains the gold standard for isotope measurements. However, they are bulky laboratory instruments. To our knowledge, only few examples exist of field applications of IRMSs. The unsurpassed creativity of Dr. Phil Ineson led him to build an IRMS mobile laboratory in a horse trailer in the early 1990s (Figure 6.2), to conduct isotopic analyses of CO_2 fluxes in the field. This mobile laboratory was applied to measure the ^{13}C:^{12}C isotope ratio of soil CO_2 fluxes in pulse-labeling experiments (Subke et al. 2009, 2012). A similar set-up was designed by Schnyder and collaborators (2004), who installed an IRMS in a heated van. The development of compact IRMS systems have facilitated the design of portable systems for automatic continuous isotope analysis in the field, such as the mobile continuous-flow IRMS system for automated field measurements of N_2 and N_2O fluxes developed by Warner et al. (2019). Besides these extraordinary cases, it is the development of laser spectroscopy that enabled high-frequency isotope analysis in the field.

Figure 6.2 *The York mobile laboratory: a mobile laboratory built in a horse trailer (a), equipped with an isotope ratio mass spectrometer (b) for the continuous measurements in the field of* ^{13}C*:*^{12}C *isotopic ratio from* ^{13}C*-enriched* CO_2 *fluxes from soil chambers (c).*

6.3 Laser absorption spectroscopy

All isotopocules (i.e., both isotopologues and isotopomers) absorb electromagnetic radiation at different wavelengths. That is, each has a characteristic absorption spectrum, with absorption being proportional to molecular density. Therefore, they can be quantified by absorption spectroscopy. However, it took several years and a lot of research for optical physicists to develop spectroscopes that had all the characteristics required to measure precise isotope ratios and δ values on gas molecules, let alone be portable and user friendly for field applications.

Generally, in absorption spectroscopy applications, the specific absorption line for each isotopocule is chosen on the basis of (1) high absorption of the line strength and (2) uniqueness, or minimal interference from other molecules in the gas sample. Additionally, the laser source needs to emit as many unique absorption lines as the isotopocules that it is intended to measure, and these need to be relatively close together, since lasers typically have a restricted tuning range. These criteria drive the choice of the particular laser source to use.

The first of such instruments used tunable diode lasers (TDLs) for the determination of $\delta^{13}C$ in CO_2 molecules, reaching a precision of 4‰ (Becker et al. 1992) (Figure 6.3), and the determination of $\delta^{13}C$ and $\delta^{2}H$ of CH_4 in air without requiring

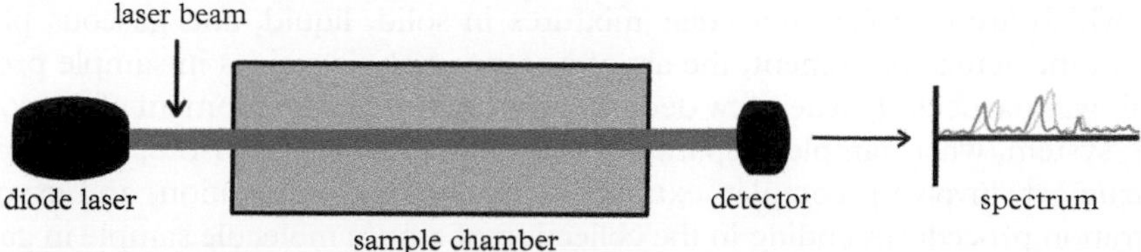

Figure 6.3 *Generalized schematic of laser spectroscopy for isotope analysis. A diode laser emits a beam that travels through a sample absorption chamber to a detector for isotope analysis. Laser spectroscopy systems vary greatly in their design and performance. We refer readers to Voglar er al. (2019) for a detailed review of different isotope analysis systems.*

sample preconcentration (Bergamaschi et al. 1994). Other spectroscopic approaches to measure C isotopes in CO_2 were also proposed, such as Fourier-transformed infrared spectroscopy (FT-IR ; Esler et al. 2000), or laser optogalvanic effect spectroscopy (Murnick and Peer 1994). But it was the use of TDLs that resulted in the first, and very exciting, continuous determination of stable C isotope ratios of ecosystem respiration in a grassland (Bowling et al. 2003). Laser spectroscopy was also successfully applied to the measurement of water isotopologues (Kerstel et al. 2002). Kerstel and collaborators (1999) developed a tunable near-infrared color center laser to make the first determination of δ^2H, $\delta^{17}O$, and $\delta^{18}O$ in water molecules, with precisions of 0.7‰ for δ^2H and 0.5‰ for $\delta^{17}O$ and $\delta^{18}O$. They later managed to improve the performance of this apparatus by using TDLs (Kerstel et al. 2002).

TDLs, however, have several significant limitations for field applications. They often use lead-salt laser diodes which require continuous cooling and therefore the entire apparatus is highly sensitive to temperature changes. TDL systems also require multiple calibration tanks and very frequent calibration (Bowling et al. 2003). For these reasons, TDL systems have been used most successfully at accessible field sites with reliable electricity on site.

Several different laser adsorption spectroscopy systems are available today, and detailed reviews of their principles of analysis and performance have been published elsewhere (Kerstel and Gianfrani 2008; Wen et al. 2012; Voglar et al. 2019). The development of cavity ring-down spectroscopy (CRDS; Berden et al. 2000), off-axis cavity-enhanced absorption spectroscopy (OA-CEAS; Baer et al. 2002), and optical feedback cavity-enhanced absorption spectroscopy (OF-CEAS ; Morville et al. 2005) have enabled very sensitive gas absorption measurements, in sturdy, portable, and easy-to-use apparatuses. These are currently the most widely used commercial lasers for isotope analysis.

6.4 Preparation systems

We have just learned that isotope analyzers at best are able to quantify isotopes in gas mixtures by laser spectroscopy, while IRMS requires pure gas samples of specific molecules. How can we measure isotopic abundances in samples used for ecological studies, which are complex molecular mixtures in solid, liquid, and gaseous phases? More than the actual instrument, the art of isotope analysis resides in sample preparation. This was particularly true a few decades ago, before the development of the conflow injection system, when sample preparation had to be done offline. Isotope analysis was then a true art, involving complex extraction, combustion, separation, and cryogenic concentration procedures ending in the collection of a pure molecule sample in gaseous phase in a sealed glass vial ready for injection through the dual inlet system (e.g. Bertolini et al. 2006).

The development of the continuous flow injection enabled coupling of other instruments, such as an EA, a thermo-combustion analyzer, and a gas or liquid chromatographer with an IRMS or lasers (Figure 6.1). Such instrument coupling revolutionized our ability to measure isotope ratios in samples of ecological and environmental interest. Sample preparation transforms the element of interest from the form in which it is present in the environmental sample to the pure molecular form suitable for IRMS or spectroscopic analysis.

Let's say we want to measure $\delta^{13}C$ and $\delta^{15}N$ in a plant tissue. To do so, we would need to convert all the organic molecules constituting the plant tissue into CO_2 and N_2, which can then be analyzed by IRMS. Which pretreatments are required? First, the solid sample would need to be combusted to create a gas mixture, and then oxidized so that all the C in the sample is converted to CO_2. Oxidation, however, will convert all the N into N_xO_y molecules, and as we learned above, an IRMS only measures N isotopes in N_2. Therefore, a reduction step following oxidation is required, to convert all N_xO_y molecules into N_2. After these two preparation steps, we have our C and N in the desired forms for isotope analysis. The sample will now be a gas mixture rich in water vapor and other molecules. A chemical trap is generally used to remove water from the sample, while gas chromatography can separate gas molecules of different species from one another, venting off the ones that are not of interest, and injecting the species of interest one at a time through the conflow injection port into the IRMS. This last step is not required in spectroscopy because spectroscopy can measure gas mixtures. Our hypothetical sample preparation can be performed by an EA connected by a continuous flow unit to an IRMS.

What if we also wanted to measure the $\delta^{18}O$ of our plant tissue sample? EA uses an external O_2 flux to combust samples. Thus, for O isotope analysis of solid samples, a high-temperature conversion EA is required because it uses pyrolysis at high temperatures to convert solid samples into gas mixtures (Figure 6.4a).

In general, air samples can be directly analyzed by laser spectroscopy, but they need to be purified and concentrated for isotope analysis by IRMS, especially when working with trace gases. Several gas chromatography and pre-concentration systems are commercially available to separate specific gas molecules from air mixtures into pure gas samples, oxidize or reduce them to obtain the desired analyte, and concentrate it if present in trace amounts, generally using cryogenic traps (Figure 6.4b).

In the case of compound-specific analysis, the preparation steps described above (i.e., oxidation, reduction, and purification of the gas molecule of interest for IRMS analysis) will need to come after the extraction and separation of the compounds of

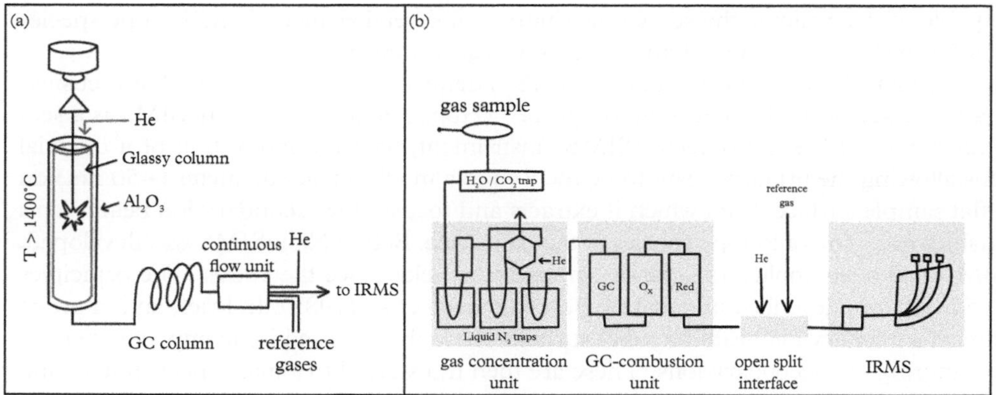

Figure 6.4 *Examples of pre-treatment systems: high temperature conversion elemental analysis (TCEA-IRMS) (a), and preconcentration with gas chromatography combustion (b).*

(a) Modified from Thermo Scientific instrument schematics. (b) Modified from Bayreuth Center of Ecology and Environmental Research Laboratory of Isotope Biogeochemistry schematic.

interest. Extractions are done offline, and several methods for the different classes of compounds of interest exist (e.g., Bligh and Dyer 1959). Most commonly, compound-specific analyses are performed by gas chromatography–combustion–mass spectrometry (GC-c-IRMS), which is now a routine method for the ^{13}C analysis of fatty acid methyl esters (e.g., Rubino et al. 2010). For O isotope analysis, thermo-combustion must be used in place of combustion. However, GC-c-IRMS can only be performed on volatile compounds. Thus, a derivatization step is required for non-volatile compounds, such as PLFAs, which are in fact methylated to fatty acid methyl esters prior to GC-c-IRMS analyses. Care must be taken to avoid errors introduced by the derivatization step, due to possible fractionations or, in the case of C isotopes, if derivative C is added in the process (Rieley 1994). Liquid chromatography (LC) is advantageous because it does not require derivatization and conversions are quantitative and do not cause fractionation. However, LC is currently only available for the determination of ^{13}C in polar thermo-labile compounds (Bodé et al. 2009). Godin and McCullagh (2011) provided an extensive review of the technical advances and limitations, as well as the possible applications of LC-IRMS. Coupling Curie-point pyrolysis with GC-c-IRMS is an alternative and powerful method to avoid offline extractions. This approach, for example, enabled the accurate quantification of natural abundance ^{13}C of specific carbohydrates, lignin, lipids, and N-containing compounds from arable soils (Gleixner et al. 1999).

6.5 NanoSIMS

All of the approaches described above enable isotope analysis of bulk solid and air samples, or at the level of specific compounds. Thus, when a team of scientists at Cameca developed the first NanoSIMS instrument (Hillion et al. 1994; Slodzian et al. 1994) capable of determining the spatial distribution of several elemental and isotope species at a 50–100 nm resolution, it appeared as a huge advancement!

From a technical standpoint NanoSIMS is defined as a "dynamic, double-focusing, magnetic-sector, multi-collecting ion probe" (Hoppe et al. 2013). NanoSIMS is a secondary ion mass spectrometry (SIMS) instrument, with the innovation of a co-axial lens allowing the primary beam to be focused to a much smaller diameter (~50 nm) on a flat sample surface, from which it extracts and focuses the secondary ion beam. This enables mapping of isotope species at the nanoscale. Before NanoSIMS was developed, SIMS had been applied in a variety of scientific fields since the 1980s. The principles of SIMS were described in detail by Benninghoven et al. (1987). In brief, an ion probe directs a primary ion beam on a flat surface where the localized ionic impact generates "sputtering" of secondary ions. These are then transferred to a mass spectrometer and counted (Figure 6.5).

In NanoSIMS, either cesium (Cs^+) or O^- are used as primary ions. The first has a smaller beam size (<50 nm) and produces negative secondary ions, while O^- has a larger beam size (~200 nm) and produces positive secondary ions (Hoppe et al. 2013). NanoSIMS can be applied to the measurement of isotopes of low and intermediate atomic number elements. H isotopes can be measured as both positive and negative secondary ions; C, O, and S isotopes are measured as negative secondary ions. N isotopes require specific conditions to be measured. In fact, positive N ions have very low

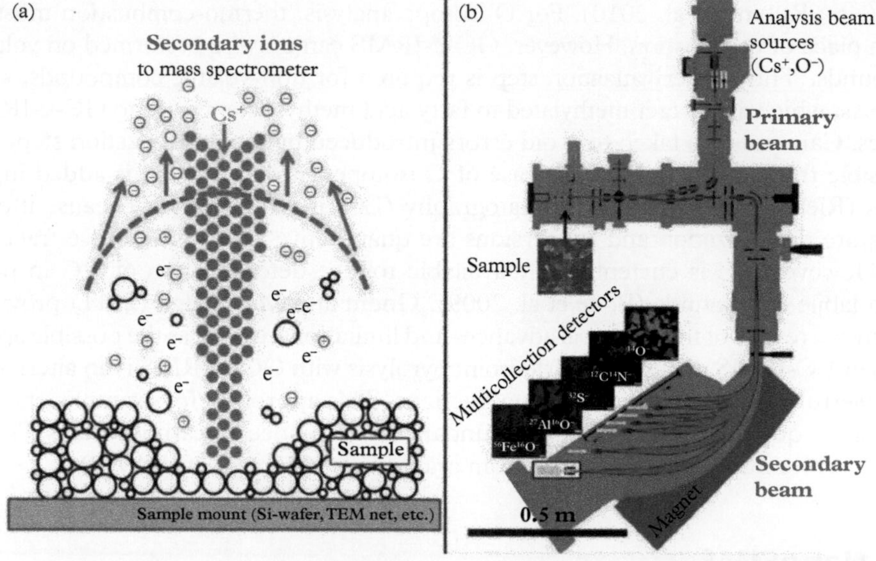

Figure 6.5 *NanoSIMS set-up showing location of primary and secondary ion beams compared to the sample surface (a). Schematic of NanoSIMS instrumentation including both primary and secondary beams (b). See online version for color detail.*
Reproduced with permission from Mueller et al. (2013) and Myrold et al. (2011).

yields and negative ions do not form. Thus, N isotopes are measured as CN^- using Cs^+ as the primary ion. Isotope measurements with NanoSIMS were reported to achieve a maximum precision of 1‰ (Hoppe et al. 2013). NanoSIMS has now been successfully applied in different fields of ecology and biogeochemistry, including studies of organo-mineral interactions on soil particles (Mueller et al. 2013) and imaging of single-cell isotopic composition at natural abundance or after stable isotope probing (Pett-Ridge and Weber 2012). In these applications, it is advisable to parallel NanoSIMS data with other spatial, textural, and compositional data from different forms of microscopy to better inform the interpretation of NanoSIMS data (Mueller et al. 2013).

6.6 A few practical hints

Analyzing the isotopic composition of a sample may provide a lot of great insight and can often be the only method for addressing a research question. However, instrumenting and staffing a single researcher's laboratory for routine isotope analysis is not a trivial endeavor.

The good news is that many stable isotope analytical facilities exist worldwide, which provide analysis on a service fee basis. Thus, scientists who do not use isotopes routinely or do not have a sample throughput large enough to justify setting up their own laboratory for isotope analysis, can still apply isotopes in their research by sending samples out for analysis.

Our suggestion is to shop carefully for the analytical facility of choice. Understanding the method used for the analysis of interest, the accuracy of measurement, the amount of sample required, the return time, and, of course, the cost should all inform the decision. Many analytical facilities have this information listed on their web page, but a call to the manager is always a good idea. We recommend identifying the facility to use at the time of the experimental design, in order to produce enough sample amount for analysis (and re-analysis, if needed), to collect and process samples within an appropriate time frame, and to set up enough replicates or level of enrichment to enable accurate isotopic determinations.

While isotope analysis is accessible to any scientist even if they are not isotope experts, we still suggest users should be well informed on the method applied for sample analysis. We recommend reviewing this chapter and the references within and consulting with instrumentation experts at isotope facilities early and often as one designs and implements studies using isotopes.

References

Baer, D. S., J. B. Paul, M. Gupta and A. O'Keefe (2002). "Sensitive absorption measurements in the near-infrared region using off-axis integrated-cavity-output spectroscopy." *Applied Physics B* 75(2): 261–265.

Becker, J. F., T. B. Sauke and M. Loewenstein (1992). "Stable isotope analysis using tunable diode laser spectroscopy." *Applied Optics* **31**(12): 1921–1927.

Benninghoven, A., F. G. Rudenauer and H. W. Werner (1987). *Secondary ion mass spectrometry: basic concepts, instrumental aspects, applications and trends*. New York, John Wiley and Sons.

Berden, G., R. Peeters and G. Meijer (2000). "Cavity ring-down spectroscopy: experimental schemes and applications." *International Reviews in Physical Chemistry* **19**(4): 565–607.

Bergamaschi, P., M. Schupp and G. W. Harris (1994). "High-precision direct measurements of $^{13}CH_4/^{12}CH_4$ and $^{12}CH_3D/^{12}CH_4$ ratios in atmospheric methane sources by means of a long-path tunable diode laser absorption spectrometer." *Applied Optics* **33**(33): 7704–7716.

Bertolini, T., I. Inglima, M. Rubino, F. Marzaioli, C. Lubritto, J.-A. Subke, A. Peressotti and M. F. Cotrufo (2006). "Sampling soil-derived CO_2 for analysis of isotopic composition: a comparison of different techniques." *Isotopes in Environmental and Health Studies* **42**(1): 57–65.

Bligh, E. G. and W. J. Dyer (1959). "A rapid method of total lipid extraction and purification." *Canadian Journal of Biochemistry and Physiology* **37**(1): 911–917.

Bodé, S., K. Denef and P. Boeckx (2009). "Development and evaluation of a high-performance liquid chromatography/isotope ratio mass spectrometry methodology for $\delta^{13}C$ analyses of amino sugars in soil." *Rapid Communications in Mass Spectrometry* **23**(16): 2519–2526.

Bowling, D. R., S. D. Sargent, B. D. Tanner and J. R. Ehleringer (2003). "Tunable diode laser absorption spectroscopy for stable isotope studies of ecosystem–atmosphere CO_2 exchange." *Agricultural and Forest Meteorology* **118**(1): 1–19.

Butler, O. T., W. R. L. Cairns, J. M. Cook, C. M. Davidson and R. Mertz-Kraus (2018). "Atomic spectrometry update—a review of advances in environmental analysis." *Journal of Analytical Atomic Spectrometry* **33**(1): 8–56.

Esler, M. B., D. W. T. Griffith, S. R. Wilson and L. P. Steele (2000). "Precision trace gas analysis by FT-IR spectroscopy. 2. The $^{13}C/^{12}C$ isotope ratio of CO_2." *Analytical Chemistry* **72**(1): 216–221.

Gleixner, G., R. Bol and J. Balesdent (1999). "Molecular insight into soil carbon turnover." *Rapid Communications in Mass Spectrometry* **13**(13): 1278–1283.

Godin, J.-P. and J. S. O. McCullagh (2011). "Review: current applications and challenges for liquid chromatography coupled to isotope ratio mass spectrometry (LC/IRMS)." *Rapid Communications in Mass Spectrometry* **25**(20): 3019–3028.

Hillion, F., B. Daigne, F. Girard, G. Slodzian, and M. Schumacher (1994). A new high performance instrument: the Cameca Nano-SIMS 50. In: A. Benninghoven, Y. Nihei, R. Shimizu and H. W. Werner, eds. *Proceedings of the Ninth International Conference on Secondary Ion Mass Spectrometry (SIMS IX), Yokohama, Japan 7–11 November 1993*. New York, John Wiley: 254–257.

Hoppe, P., S. Cohen and A. Meibom (2013). "NanoSIMS: technical aspects and applications in cosmochemistry and biological geochemistry." *Geostandards and Geoanalytical Research* **37**(2): 111–154.

Kerstel, E. and L. Gianfrani (2008). "Advances in laser-based isotope ratio measurements: selected applications." *Applied Physics B* **92**(3): 439–449.

Kerstel, E. R. T., G. Gagliardi, L. Gianfrani, H. A. J. Meijer, R. van Trigt and R. Ramaker (2002). "Determination of the $^2H/^1H$, $^{17}O/^{16}O$, and $^{18}O/^{16}O$ isotope ratios in water by means of tunable diode laser spectroscopy at 1.39 μm." *Spectrochimica Acta Part A: Molecular and Biomolecular Spectroscopy* **58**(11): 2389–2396.

Kerstel, E. R. T., R. van Trigt, J. Reuss and H. A. J. Meijer (1999). "Simultaneous determination of the $^2H/^1H$, $^{17}O/^{16}O$, and $^{18}O/^{16}O$ isotope abundance ratios in water by means of laser spectrometry." *Analytical Chemistry* **71**(23): 5297–5303.

Lynch, L. M., M. B. Machmuller, M. F. Cotrufo, E. A. Paul and M. D. Wallenstein (2018). "Tracking the fate of fresh carbon in the Arctic tundra: will shrub expansion alter responses of soil organic matter to warming?" *Soil Biology and Biochemistry* **120**: 134–144.

McKinney, C. R., J. M. McCrea, S. Epstein, H. A. Allen and H. C. Urey (1950). "Improvements in mass spectrometers for the measurement of small differences in isotope abundance ratios." *Review of Scientific Instruments* **21**: 724–730.

Morville, J., S., Kassi, M. Chenevier and D. Romanini (2005). "Fast, low-noise, mode-by-mode, cavity-enhanced absorption spectroscopy by diode-laser self-locking." *Applied Physics B* **80**: 1027–1038.

Mueller, C. W., A. Kölbl, C. Hoeschen, F. Hillion, K. Heister, A. M. Herrmann and I. Kögel-Knabner (2012). "Submicron scale imaging of soil organic matter dynamics using NanoSIMS—from single particles to intact aggregates." *Organic Geochemistry* **42**(12): 1476–1488.

Mueller, C. W., P. K. Weber, M. R. Kilburn, C. Hoeschen, M. Kleber and J. Pett-Ridge (2013). Advances in the analysis of biogeochemical interfaces: NanoSIMS to investigate soil microenvironments. In: D. L. Sparks, ed. *Advances in agronomy*, Vol. 121. New York, Academic Press: 1–46.

Murnick, D. E. and B. J. Peer (1994). "Laser-based analysis of carbon isotope ratios." *Science* **263**(5149): 945.

Myrold, D. D., J. Pett-Ridge and P. J. Bottomley (2011). "Nitrogen mineralization and assimilation at millimeter scales." *Methods in Enzymology* **496**(3): 91–114.

Nier, A. O. (1940). "A mass spectrometer for routine isotope abundance measurements." *Review of Scientific Instruments* **11**(7): 212–216.

Pett-Ridge, J. and P. K. Weber (2012). "NanoSIP: NanoSIMS applications for microbial biology." *Methods in Molecular Biology (Clifton, N.J.)* **881**: 375–408.

Rieley, G. (1994). "Derivatization of organic compounds prior to gas chromatographic–combustion–isotope ratio mass spectrometric analysis: identification of isotope fractionation processes." *Analyst* **119**(5): 915–919.

Rubino, M., J. A. J. Dungait, R. P. Evershed, T. Bertolini, P. De Angelis, A. D'Onofrio, A. Lagomarsino, C. Lubritto, A. Merola, F. Terrasi and M. F. Cotrufo (2010). "Carbon input belowground is the major C flux contributing to leaf litter mass loss: evidences from a ^{13}C labelled-leaf litter experiment." *Soil Biology and Biochemistry* **42**(7): 1009–1016.

Schnyder, H., R. Schäufele and R. Wenzel (2004). "Mobile, outdoor continuous-flow isotope-ratio mass spectrometer system for automated high-frequency ^{13}C- and ^{18}O-CO_2 analysis for Keeling plot applications." *Rapid Communications in Mass Spectrometry* 18(24): 3068–3074.

Slodzian, G., B. Daigne, F. Girard and F. Hillion (1994). Ion optics for a high resolution scanning ion microscope and spectrometer: transmission evaluations. In: A. Benninghoven, Y. Nihei, R. Shimizu and H. W. Werner, eds. *Proceedings of the Ninth International Conference on Secondary Ion Mass Spectrometry (SIMS IX), Yokohama, Japan 7–11 November 1993*. New York, John Wiley: 294–297.

Subke, J.-A., A. Heinemeyer, H. W. Vallack, V. Leronni, R. Baxter and P. Ineson (2012). "Fast assimilate turnover revealed by in situ $^{13}CO_2$ pulse-labelling in Subarctic tundra." *Polar Biology* **35**(8): 1209–1219.

Subke, J.-A., H. W. Vallack, T. Magnusson, S. G. Keel, D. B. Metcalfe, P. Högberg and P. Ineson (2009). "Short-term dynamics of abiotic and biotic soil $^{13}CO_2$ effluxes after in situ $^{13}CO_2$ pulse labelling of a boreal pine forest." *New Phytologist* **183**(2): 349–357.

Voglar, G. E., S. Zavadlav, T. Levanič and M. Ferlan (2019). "Measuring techniques for concentration and stable isotopologues of CO_2 in a terrestrial ecosystem: a review." *Earth-Science Reviews* **199**: 102978.

Warner, D. I., C. Scheer, J. Friedl, D. W. Rowlings, C. Brunk and P. R. Grace (2019). "Mobile continuous-flow isotope-ratio mass spectrometer system for automated measurements of N_2 and N_2O fluxes in fertilized cropping systems." *Scientific Reports* **9**(1): 11097.

Wen, X.-F., X. Lee, X.-M. Sun, J.-L. Wang, Y.-K. Tang, S.-G. Li and G.-R. Yu (2012). "Intercomparison of four commercial analyzers for water vapor isotope measurement." *Journal of Atmospheric and Oceanic Technology* **29**(2): 235–247.

Activity: Developing a Stable Isotope Measurement Plan

Objectives

- Practice developing a plan for measuring stable isotopes in ecological samples.
- Compare capabilities and pricing at isotope analytical facilities accessible to you.

Tools

This activity only requires an internet connection to research isotope analytical facilities. Refer to the information regarding capabilities, limitations, and considerations for different stable isotope quantification methods in Chapter 6.

Exercise 1

Develop a stable isotope measurement plan

You find yourself in a situation where you need to measure stable isotope natural abundance on the ecological samples described in Table A6.1 to address your research question. Assuming you do not have the instrumentation needed in your own laboratory, develop a plan for measuring stable isotopes in your samples. Research isotope analytical facilities. Your plan should include the following:

- Instruments required to measure stable isotopes in the samples of interest.
- Identify an isotope analytical facility with the instruments needed.
- Notes on sample preparation prior to shipping to the facility.
- Itemized budget for the total cost of analysis.

Table A6.1 *Ecological samples collected for stable isotope analysis. Values given are realistic but were made up for the purpose of this activity*

Sample type	Amount of sample	Number of samples	Additional information	Stable isotopes to be measured
Soil	1 g	8	1% C, 0.1% N	$\delta^{13}C$, $\delta^{15}N$
Air	20 mL	8	415 ppm CO_2	$\delta^{13}C$, $\delta^{18}O$
Water	100 mL	4		$\delta^{18}O$, $\delta^{2}H$

Review questions

1. What are the benefits and limitations of the different types of analytical approaches (i.e., IRMS versus laser spectroscopy)?
2. Consider scenarios in which you might measure stable isotopes during your own research. Which measurement techniques would be most appropriate for your research?
3. How many samples would you need to routinely analyze to justify purchasing your own instrument versus sending samples out to service laboratories? Consider sample throughputs for both IRMS and laser spectroscopy.

Index